THE FORTH

—— A PICTURE HISTORY ——

THE FORTH BRIDGE

A PICTURE HISTORY

BIRLINN

Opposite: On the left of the group is Benjamin Baker, the designer, and on the right, William Arrol, 2 March 1888. Visitors were encouraged to visit the site, and as Wilhelm Westhofen noted in The Forth Bridge *(1890): 'As in most other matters ladies were to the fore, pluckily climbing into every nook and corner where anything interesting might be seen or learned, up the hoists and down the stairs and ladders, and frequently leaving the members of the so-called stronger sex far behind.'*

© Sheila Mackay, 1990
First published by Moubray House Publishing,
Edinburgh, in 1990
Published by HMSO in 1993
Published by Mercat Press, Edinburgh, in 2001
Reprinted by Mercat in 2003 and 2006

This edition published in 2011 and reprinted
in 2013 and 2017 by

Birlinn Ltd
West Newington House
10 Newington Road
Edinburgh
EH9 1QS

www.birlinn.co.uk

ISBN: 978 1 84158 935 0

Designed by Dorothy Steedman at Lackie Newton,
Edinburgh

Typeset in Century Nova by Artwork Associates,
Edinburgh

Printed in China through World Print Ltd

The author wishes to thank Richard Packer, photography supervisor in the Civil Engineering Department at Imperial College, London, for his enthusiasm and for the care he took in making the prints used in this book; Andrew Patrizio of the Talbot Rice Gallery, Edinburgh, for information on the work of Evelyn Carey; Miles Oglethorpe and Geoffrey Stell of the Royal Commission on the Ancient and Historical Monuments of Scotland; and Douglas McBeth of Kenchington, Little & Partners, Engineers, Edinburgh.

The photographs in this book are taken from prints held in the library of the Civil Engineering Deparment, Imperial College of Science, Technology and Medicine, London, and are reproduced by kind permission of the departmental librarian. The publisher is grateful to the Royal Commission on the Ancient and Historical Monuments of Scotland, which commissioned the photographic reproductions of the Imperial College prints used in this publication. These prints are now held in the National Monuments Record of Scotland.

The Publisher is also grateful to the National Library of Scotland for providing the prints of visitors to the Bridge and the Opening Ceremony, taken from contemporary editions of the *The Illustrated London News*. All other drawings in this publication are taken from Wilhelm Westhofen's *The Forth Bridge*.

CONTENTS

The Forth Bridge from the north side, 23 May 1887.

'The Engineers with their gigantic works sweep everything before them in this Victorian era,' wrote the Forth Bridge's designer, Benjamin Baker. In the 1880s the east coast line cut an audacious swathe from Dover, via London and York to Edinburgh, but the Forth estuary remained unbridged. Baker's inspired cantilever design met the challenge: the Forth Bridge would be the greatest engineering feat the world had ever seen, akin to a moonlanding in our own times, the largest railway bridge ever built and the first with a steel superstructure. The impact of huge-scale industry on the peaceful hamlets of North and South Queensferry was staggering, making houses, ships and people seem like mere toys 'as the mighty piers slowly arose out of the sea and the ascending columns climbed ever higher and higher'. In this photograph taken in Jubilee Year the three cantilever columns and the railway viaduct, supported by masonry piers, have reached their final height. The cantilever arms were then built outwards and upwards from the foundations and outwards and downwards from the top of the towers to be linked by girder arms.

THE FORTH BRIDGE.
FROM THE N. SIDE.

'The Bridge'

The Forth Bridge was the bridge of dreams: the largest ever built and the first with a steel superstructure. 'A long stride over space . . . the longest distance between supports yet covered by mechanical means,' said *The Illustrated London News* in 1889 of the greatest of all that century's great engineering adventures. 'A romantic chapter from a fairytale of science,' said the bridge's designer, Benjamin Baker.

'The bridge is a style unto itself: the simple directness of purpose with which it does its work is splendid and invests your vast monument with a kind of beauty of its own, differing though it certainly does from all the beautiful things I have ever seen,' said the architect Alfred Waterhouse. One hundred years later his tribute to Scotland's Forth Bridge stirs affectionate recognition. And through the long lens of history we become aware that there was a decade one century ago when Scots achieved something of the order of a moon landing on their own doorstep.

So it was not surprising that, when industry invaded the peaceful hamlets of North and South Queensferry, and thousands of men from all over Europe arrived to labour for seven years, day and night (Sundays excluded), people rubbed their eyes in disbelief. For them, 'The Bridge' would be as momentous an achievement as space exploration in our own times.

The Forth Bridge will be 'a triumph of engineering skill [to] eclipse the Ship Canal which has turned Africa into an island, reduce the Pyramids to mere child's play and, in all likelihood, lead to a revolution in the art of constructing bridges of this description,' declared a newspaper of the time.

The time was ripe. Like all other world wonders (it was said to be the eighth in modern times) the Forth Bridge was born of a mixture of necessity and technological daring. 'The Engineers with their gigantic works sweep everything before

them in this Victorian era,' wrote Benjamin Baker of his colleagues who pitted imagination and invention against every conceivable natural obstacle to construct a bridge or a tunnel, a cutting or an embankment.

They were virtually unstoppable. In the 1880s it was possible for a Victorian to travel from Dover via London, York and Newcastle to Edinburgh in fine style. But when the great locomotive juddered to a halt near Queensferry and the sound of steam died away, the only sound to be heard was the lapping water of the Forth as he waited for the ferry and prayed for a good crossing.

The east coast railway line cut an audacious swathe through the British countryside from Dover to the Forth, from Fife to the Tay and from Dundee to Aberdeen. However, the so-far untamed deep-sea inlets of the Rivers Tay and Forth still stubbornly flaunted Nature's supremacy over Progress.

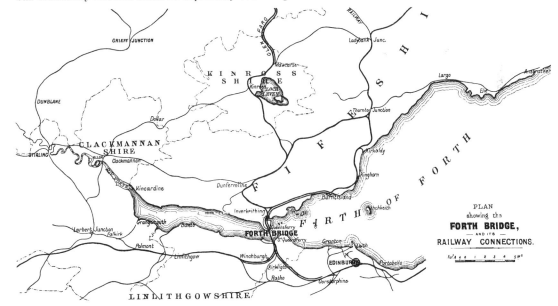

PLAN
showing the
FORTH BRIDGE,
— AND ITS —
RAILWAY CONNECTIONS.

Opposite: A Forth Bridge cantilever compared to famous buildings of the world. This illustration, drawn by Professor Cockerell, first appeared as a wall diagram at the South Kensington Museum and has been reproduced widely since. It shows a cantilever compared in size with the Great Pyramid at Giza, St Peter's, Rome, St Paul's, London, Chartres Cathedral, the Central Transept at Crystal Palace, etc.

In his 1887 lecture to the Edinburgh Literary Institute, Benjamin Baker used city landmarks to convey the enormous size of the Forth Bridge: 'To get an idea of the spans, let them stand on Waverley Bridge and look towards the Castle, and consider that the engineers had to span that distance with a complicated structure weighing 150,000 tons without the possibility of any intermediate pier or support; and let them consider also that the rail level would be as high above the sea as the Castle esplanade was above Princes Street, and that the steelwork of the bridge would soar 200 feet higher.' For a London audience the same year, Baker used London landmarks: Green Park, Piccadilly and St Paul's Cathedral.

The Tay Bridge disaster of 1879 had seemed to confirm Nature's supremacy. Eager to exploit the troubles of the east coast railway companies, the wily west coast barons stepped up their schemes to divert custom away through miles of new railway links, to the west and north to the Highlands and the vast sporting estates. But their scheming was squashed in the nick of time by a far more ambitious plan. The Tay Bridge would be reconstructed with William Arrol as contractor and later he would also take on the colossal bridge whose design was being thoroughly reconsidered by the Forth Bridge Railway Company following the tragedy on the Tay.

The builders of the Forth Bridge took extraordinary precautions to allay public anxiety which lingered long after the collapse of the Tay Bridge. Stringent requirements were set for the new design, which, with hindsight, was said to be over-designed. This bridge will 'by its freedom from vibration gain the confidence of the public, and enjoy the reputation of being not only the biggest and strongest, but also the stiffest bridge in the world,' reassured Wilhelm Westhofen, the supervisor of the Inchgarvie (middle) cantilever.

People watched in amazement as, from 1882, staggering amounts of materials began to arrive on both shores of the Forth by sea and rail. Industry strode confidently through the sea-side towns, first in the shape of gigantic mansonry pillars to support the railway viaduct, so gigantic that houses, ships and people seemed reduced to the scale of mere toys in a child's game. Astonishingly, the bridge was virtually built twice: each part of the structure was put together on land and disassembled before being sent out to the bridge site for final erection.

The noise ringing out in the clear air was incredible, of steel hitting steel, of hammers, riveters, hydraulic spades and generators, as men worked everywhere, on the beaches, in the drill roads and the hastily constructed workshops.

There was, of course, a sacrifice to be made. Fifty-seven lives were lost and 461 men, whose fate is unrecorded, were taken to the Edinburgh Hospital. 'It is impossible to carry out a gigantic work without paying for it, not merely in money, but in men's lives,' said Benjamin Baker. Grim though that toll seems, given the dangers of working below sea level and at great heights, the engineers and contractors were relieved that it was not even higher.

'From year to year the wonder grew as the mighty piers [cantilevers] slowly arose out of the sea and the ascending columns climbed ever higher and higher,' wrote William Arrol's biographer. Fortunately, he was not alone in recording progress. The construction of the Forth Bridge was documented as assiduously as this century's moon landings by newspapers and magazines up and down the land, providing readers with almost blow-by-blow accounts of events at Queensferry from the start of work in 1883 until the opening ceremony in March 1890.

But the records of two men, both engineers on the bridge, stand out. Wilhelm Westhofen's book *The Forth Bridge* (1890) and Evelyn Carey's fine series of glass plate negatives (many of which have not been published before) comprise the most comprehensive account of any of the great engineering feats of the nineteenth century. Carey's photographs, executed with technological brilliance in the infancy of photography, is particularly impressive when combined with Westhofen's incisive commentary. Their work is the inspiration for this book which celebrates the *grande dame* of bridges in her centenary year.

During construction Wilhelm Westhofen estimated that the bridge site was visited by 'one-tenth of all people distinguished by rank or by scientific or social attainment' with women 'to the fore'. And for several decades after the opening in March 1890, the dazzling achievement thrust Scotland into the forefront of engineering enterprise. But soon the motoring age dawned and men began to dream another dream – of a road bridge over the Forth, an alternative to waiting for the car ferry with dread of a 'bad crossing' or driving the long way round via the Kincardine Bridge.

With the passing of the steam age and the arrival of the upstart new road bridge, the Forth Bridge lost some of her glamour. Even Scots who threw coins out of carriage windows into the sea for luck countless times throughout childhood began to take her for granted. 'The Bridge' seemed like an old-fashioned ornament on a cupboard shelf, still useful and too laden with memories to throw out, but somehow too incongruous to fit easily into the computer age. She was built for steam trains. The sealed carriage windows of Intercity 125s discourage penny-throwing as we glide rather than chug across to Fife and back.

But the rest of the world continues to beat a path to the Forth. The Bridge is as synonymous with Scotland as tartan, heather and haggis. Thousands come every year from Europe, America, Japan and all over Britain to ponder the vast red monument. As Benjamin Baker said of the only structure like it anywhere: 'the Eiffel Tower is a foolish piece of work, ugly, ill-proportioned and of no real use to anyone.' He was right. The Forth Bridge is more than three times the size, was built across the sea and continues to do a proper job of work. For Scots everywhere 'The Bridge' will always be an integral part of life's landscape, affectionately honoured and proudly celebrated as she lumbers into her second century.

James Anderson's design for a bridge over the Forth, 1818.
In The Forth Bridge (1890) Wilhelm Westhofen commented that it was 'so light indeed that on a dull day it would hardly have been visible and, after a heavy gale, probably no longer to be seen on a clear day either'.

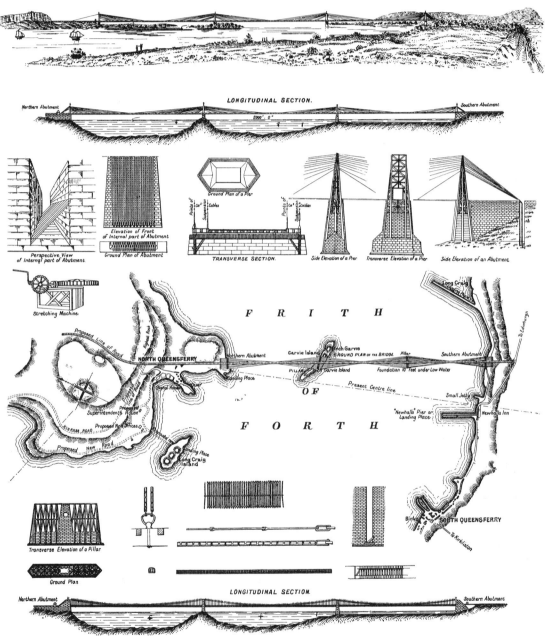

FIG. 1. ANDERSON'S DESIGN FOR BRIDGE OVER THE FORTH, 1818.

One day there would be a bridge. Man would one day have the power to fly, to travel to the moon, to bridge the Forth Estuary. Thoughts like these must have coloured the dreams of travellers in Scotland who had experienced 'bad crossings' of the temperamental Forth estuary by ferry or wasted days taking the long way round via Stirling on foot or horseback. As Wilhelm Westhofen records in *The Forth Bridge:* 'The Roman leader bent on exploration and conquest, probably conjured up in his mind's eye the faint outlines of a bridge as he trudged the weary miles along the south shore and found neither boats to carry him across nor ford to traverse; so must the sainted Margaret, the wife of Malcolm Caen-Mohr, on her frequent pious pilgrimages between Edinburgh, Linlithgow and Dunfermline about the time of the Norman Conquest, and so probably her son Alexander the First of Scotland, who in attempting to cross from South to North Queensferry was overtaken by a gale and beaten down the Firth, and had finally to land on the island of Inchcolm, five miles away ... So must also many of the poor wayfarers who got drenched to the skin and suffered the horrors of sea-sickness during the crossing.'

Schemes for bridging the Forth estuary at the Queensferry Passage were discussed and abandoned throughout the nineteenth century. The success of the Rotherhithe Tunnel under the River Thames even led to surveys for a Forth Tunnel. In 1806 a prospectus was issued by 'a number of noblemen and gentlemen of the first respectability and scientific character' which invited the public to subscribe to a double tunnel, 'one for comers and one for goers'. Nothing came of the project and the next scheme to be seriously considered was submitted in 1818 by James Anderson, an Edinburgh civil engineer and surveyor. His 'Design for a Chain Bridge' was the irresistible target of Wilhelm

Westhofen's wit: 'To judge by the estimate, the designer can hardly have intended to put more than from 2000 to 2500 tons of iron into the bridge, and this quantity distributed over the length would have given the structure a very light and slender appearance, so light indeed that on a dull day it would hardly have been visible, and after a heavy gale probably no longer to be seen on a clear day either'. Thankfully that scheme floundered too, and, says Westhofen, 'for forty years more the travelling public put up with what they could get'. Some did not make it even to the ferry-boats. An 'unfortunate party of people were driving down the Hawes Brae at so rapid a pace that horses, carriage and passengers went right off the pier into the water. and none of them came out alive'.

The next phase in the story of the search for the bridge of dreams hinged dramatically on the public disaster of the Tay Bridge and the private nightmare of Sir Thomas Bouch who was considered a force to be reckoned with in the age of Railwaymania. In 1849 the Edinburgh and Northern Railway changed its name to the Edinburgh, Perth and Dundee Railway and recruited the young Thomas Bouch as its engineer. He was determined to come up with an answer to crossing the Tay and Forth estuaries and devised a system of transporting trains over water on floating platforms. His idea was not entirely original, nevertheless his name became associated with these 'floating railways' which were considered to be marvels of ingenuity and invention, so much so that the directors of the North British Railway Company took his plans to bridge the two estuaries very seriously indeed.

Thomas Bouch's 20-year-old dream assumed reality in July 1871 with the laying of the foundation stone of the Tay Bridge. And by December 1879 the Bouch family, ensconced in their comfortable Edinburgh West End home, had more reason to count their blessings than to worry about the storm raging over Scotland's east coast. The Tay Bridge, for which Thomas Bouch was knighted, had been in operation for 19 months; Margaret Bouch had laid the foundation stone for the Forth Bridge in 1873; the Bouch's engineering son was supervising his father's designs for the bridge over the South Esk at Montrose.

Work was under way on the Forth Bridge which was to be the head of the household's ultimate achievement. Together, the three bridges would complete the long-dreamed-of route to the north, from Edinburgh to Dundee and Aberdeen, carrying the trains of the railway companies over the deep-sea inlet barriers of the east coast at last.

The Forth Bridge Company was formed in 1873 to carry out Thomas Bouch's design for a suspension bridge and the same year an Act of Parliament authorised construction and contracts were signed with Messrs W. Arrol and Co., of Glasgow. The plans looked so promising that, at the zenith of his career, Sir Thomas little expected the blow which arrived at his home in the form of a telegram three days after Christmas 1879:

Terrible accident on bridge one or more of highgirders blown down a.m. not sure of the safety of the last down Edinbr train will advise further as soon as can be obtained

The extent of the disaster was all too evident by the cruel light of dawn. The Tay Bridge had crumbled into the estuary, sweeping the Edinburgh train and an estimated 75 passengers to their deaths with the debris of brick and metal. Work on the Forth was immediately halted by a formal Abandonment Act pending investigations into the disaster. Public confidence in Sir Thomas and his assumed abilities was shaken to the core. His 'floating railways' continued to operate and locomotives went on chugging over the 300 miles of lines he constructed in Scotland and England. But the Tay Bridge collapse rendered these achievements virtually worthless and Sir Thomas retired shattered to his country house in Moffat where he died on 1 November 1880 from a cold which he had no will to resist. He was buried at Edinburgh's Dean Cemetery.

It would, however, have taken more than a disaster even of this scale to halt a determined Victorian railway director in his tracks. Before 1880 was out the four railway companies which had raised the capital for Bouch's scheme – the Great Northern, the North-Eastern, the Midland and the North British – had instructed their consulting engineers, Messrs Barlow, Harrison and Fowler to reconsider all the options for bridging the Forth.

A design submitted by John Fowler and Benjamin Baker, based on the cantilever and central girder principle, was modified to take account of the conflicting views of the consultant engineers and submitted to the director of the Forth Bridge Company in May 1881. An Act of Parliament passed in July 1882 authorised construction of the new design with the stipulation that the Board of Trade should maintain an overall inspection of the work during construction.

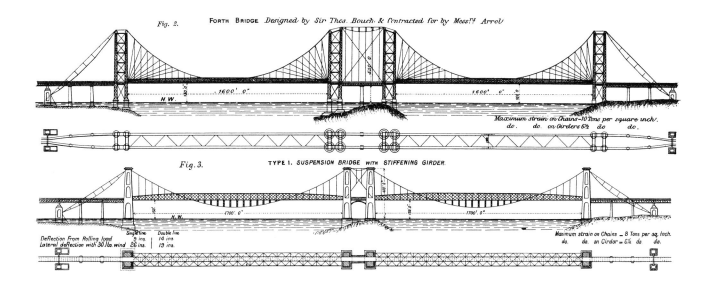

Fig. 2.   FORTH BRIDGE *Designed by Sir Thos. Bouch & Contracted for by Messrs Arrol*

H.W.

1600' 0"   1600' 0"

Maximum strain on Chains–10 Tons per square inch.
do.   do.   on Girders 6½   do.   do.

Fig. 3.   TYPE 1. SUSPENSION BRIDGE WITH STIFFENING GIRDER.

H.W.

1700' 0"   1700' 0"

	Single line	Double line
Deflection From Rolling load	9 ins.	14 ins.
Lateral deflection with 30 lbs. wind	26 ins.	19 ins.

Maximum strain on Chains – 8 Tons per sq. inch.
do.   do.   on Girder 6¼   do.   do.

Overpage: The original and the final design for a Forth Bridge based on the cantilever principle with linking girder arms submitted by John Fowler and Benjamin Baker. The design was accepted in 1881 after the railway companies and their consulting engineers had scrutinised several options, including schemes for tunnels and bridges with tunnels, in the aftermath of the Tay Bridge disaster.

Alternative designs for the Forth Bridge. Preparatory work on Thomas Bouch's design (1865) for a suspension bridge carrying two lines of rails (top) was halted following the Tay Bridge disaster in December 1879.

The Directors of the Forth Bridge Railway Company, 15 April 1884.
The directorate was made up of the chairman and vice-chairman of the four interested companies: the Midland, Great Northern, North-Eastern and North British Railway Companies.

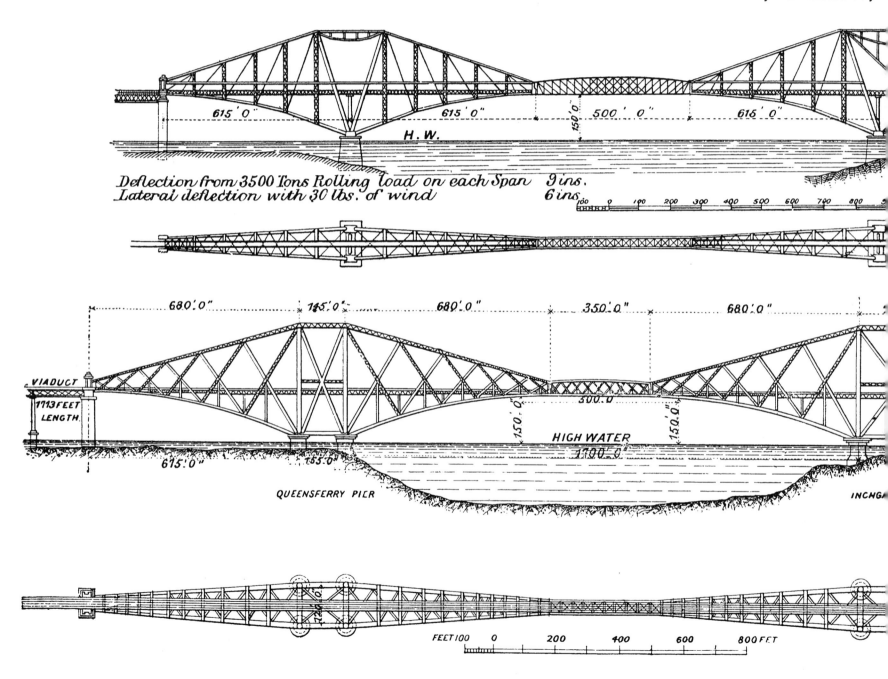

615' 0"   615' 0"   150' 0"   500' 0"   616' 0"

H.W.

Deflection from 3500 Tons Rolling load on each Span   9 ins.
Lateral deflection with 30 lbs. of wind   6 ins.

100   0   100   200   300   400   500   600   700   800

680' 0"   145' 0"   680' 0"   350' 0"   680' 0"

VIADUCT

1713 FEET
LENGTH

150' 0"   500' 0"   150' 0"

HIGH WATER

1100' 0"

615' 0"   155' 0"

QUEENSFERRY PIER   INCHGA

120' 0"

FEET 100   0   200   400   600   800 FEET

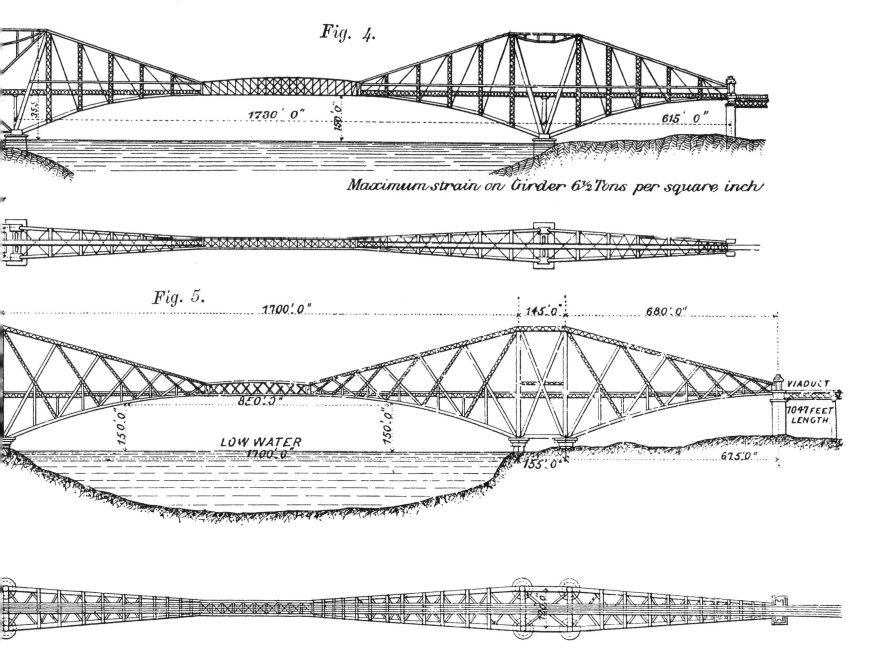

Fig. 4.

1730' 0"   150' 0"   615' 0"

Maximum strain on Girder 6½ Tons per square inch

Fig. 5.

1700' 0"   145' 0"   680' 0"

850' 0"

150' 0"   150' 0"

LOW WATER
1700' 0"

155' 0"   675' 0"

VIADUCT

7047 FEET LENGTH

The Human Cantilever was devised by Benjamin Baker to explain the cantilever principle at his Royal Institution lecture of 1887. The man in the centre was Kaichi Watanabe, one of the first generation of Japanese engineers sent to study, and ultimately imitate and surpass, western technical achievements. Watanabe, a student of John Fowler and Benjamin Baker, was invited to participate in the human model of the cantilever to remind audiences of the debt the designers owed to the Far East where the cantilever principle was invented. Fowler and Baker's design for the Forth Bridge was revolutionary because of the linking girder arms and its unprecedented colossal size. 'Two men sitting on chairs extend their arms and support the same by grasping sticks which are butted against the chairs,' explained Baker. 'There are thus two complete piers, as represented in the outline drawing above their heads. The centre girder is represented by a stick suspended or slung from the two inner hands of the men, while the anchorage provided by the counterpoise in the cantilever end piers is represented here by a pile of bricks at each end. When a load is put on the central girder by a person sitting on it, the men's arms and the anchorage ropes come into tension, and the men's bodies from the shoulders downwards and the sticks come into compression. The chairs are representative of the circular granite piers. Imagine the chairs one-third of a mile apart and the men's heads as high as the cross of St Paul's, their arms represented by huge lattice steel girders and the sticks by tubes 12 feet in diameter at the base, and a very good notion of the structure is obtained.'

A tremendous leap of faith inspired the design of the Forth Bridge which was to become the greatest engineering feat the world had ever seen, the largest railway bridge ever built and the first with a steel superstructure.

Of all the adventures of the engineering entrepreneurs this would be the greatest. Benjamin Baker was well aware of the momentous challenge before him when he wrote: 'If I were to pretend that the designing and building of the Forth Bridge was not a source of present and future anxiety to all concerned, no engineer of experience would believe me. Where no precedent exists, the successful engineer is he who makes the fewest mistakes.'

In 1867 he had published a series of articles in *The Engineer* advocating the use of cantilevers supporting a girder system as the most effective means of constructing long-span bridges. Considering the man he was, it is easy to imagine his real sense of humility at the twists of fate which 15 years later, after the abandonment of Bouch's Forth Bridge scheme, gave him the chance to convert his theory to reality on a gigantic stage.

The use of the cantilevers with linking girder arms to bridge the Forth must have seemed radical at the time but, as *The Illustrated London News* observed in 1889: 'the germ of the idea can be traced back to the earliest form of architecture . . . Many of our readers may have visited Cashmere and noticed the primitive bridges over the Jhelum in that locality. These bridges like the Forth Bridge have more than one span . . . Sir John Fowler and Mr Baker are perfectly well acquainted with these primitive types; but their high merit consists in having the comprehensive grasp of the idea which enabled them to see how this early form of construction could be applied to the requirements of modern engineering.'

Wilhelm Westhofen was at pains to dispel any thoughts that the cantilever and girder method of construction might

be 'a modern and patentable invention'. 'As a matter of fact, it is a pre-historic arrangement. In the earliest Egyptian and Indian temples will be found the stone corbel and lintel combinations and in the oldest, as in the most modern wooden bridges will be seen practically the same thing . . . Skeleton bridges on a similar principle have for ages past been thrown by savages across rivers. Perhaps one of the most interesting structures of this kind ever built is a bridge in Thibet, constructed about 220 years ago.'

Fowler and Baker submitted designs for steel cantilever bridges across the River Severn in 1864 and 1871 and Baker designed a superstructure for a proposed ferry bridge across the River Tees in 1873. A 148 ft railway bridge built in 1876 on the cantilever system over the River Warthe near Posen was the only precedent when work began on the Forth Bridge, and that was no match for the mighty structure planned by Baker and Fowler.

Timber bridge from The Forth Bridge *chosen by Westhofen to illustrate primitive cantilever bridges. 'Perhaps one of the most interesting structures of this kind ever built is a bridge in Thibet, constructed about 220 years ago.'*

Inchgarvie 12 ft. tube junction on the drill roads, 2 March 1888.
Benjamin Baker the designer (left) and William Arrol the contractor (right) with two workmen (above) and, standing below Arrol, one of the women who, as Wilhelm Westhofen noted in The Forth Bridge, *were 'to the fore, pluckily climbing into every nook and corner where anything interesting might be seen or learned'.*

'Of the present design it may be truly said that all anticipations have been most brilliantly realised,' said Westhofen. All too aware that public anxiety in the wake of the Tay Bridge disaster was still running high, stringent requirements were set for the new design. 'We think it will be apparent to most engineers and bridge builders that the original suspension-bridge design (Thomas Bouch's) complied with none of these conditions whilst the girder design complies with all.'

The largest railway bridge ever built would 'by its freedom from vibration gain the confidence of the public, and enjoy the reputation of being not only the biggest and strongest, but also the stiffest bridge in the world'; at every stage of erection the incomplete structure would be 'as secure against a hurricane as the finished bridge'.

Cast iron, used as a structural material on the Tay Bridge, was rejected in favour of steel. The Forth Bridge would be the first large structure made of steel. Intent on exploiting the properties of this relatively new material to the full, Fowler and Baker nevertheless retained caution. The structural steel would comply with specifications laid down by the Admiralty, Lloyd's and the Underwriters' Registry 'as determined by the experience gained in the use of many thousands of tons of steel plates, bars and angles for shipbuilding purposes'.

THE BRIDGE BUILDERS

Benjamin Baker 1840-1907
The Designer

'Not by any means a big man; but a very great one'

Benjamin Baker was particularly well-qualified for the task before him. He combined a flair for engineering design with an understanding of the relevant materials which he had gained in the most practical way – not through academic schooling so much as an apprenticeship served in the famous South Wales Ironworks at Neath Abbey which he entered at the age of 16.

After completing his training, he moved to London where he worked under W. H. Wilson on the construction of Victoria Station before joining the firm of John Fowler in 1862. He became a partner in 1875 and remained associated with the firm until Fowler's death in 1898 when he assumed responsibility for much of the work of the business.

Apart from the Forth Bridge, the work undertaken during these years was of epic proportions, even by the standards of Victorian engineering achievement. It is to Fowler and Baker, more than anyone else, that London owes its Underground Railway Network, which began as Fowler's Metropolitan Railway in the early 1860s and developed with Baker's pioneering work on the 'tube' system in the 1890s. Baker also acted as consulting engineer on the Aswan Dam (1894-1902) and assisted John Dixon in 1878 in the perilous business of transporting the great monolithic obelisk called Cleopatra's Needle from Egypt to Britain by sea.

He was remembered in old age by an employee: 'In the office of Baker and Hutzig at Queen Square Place, I first saw, with his magnificent head, firm jaw, generous moustache and penetrating and fearless eyes; a most impressive figure, not by any means a big man; but a very great one'. Benjamin Baker was knighted for his work on the Forth Bridge in 1890.

Sir John Fowler the consulting engineer (left) with Lady Fowler and William Arrol (undated photograph).

William Arrol

John Fowler

Benjamin Baker

William Arrol 1839-1913
The Contractor

'The very stuff of Victorian dreams'

In the mid-1880s, when he was acting as contractor on the new Tay Bridge as well as the Forth Bridge, William Arrol rose at five each Monday morning in his Glasgow home to be at his nearby Dalmarnock works at six to inspect progress. He would then board the train for Edinburgh and the Forth Bridge, where he would work for the rest of Monday and Tuesday before catching the Dundee train to arrive at the Tay Bridge at about 11 pm. At six the following morning he would be at work there, returning late at night to Glasgow to be back at his works first thing the following morning.

Then he was off again to the Forth and the Tay before catching the London train where, on Saturday, he would spend the day discussing progress at the offices of Fowler and Baker. Unless business in London was incomplete (in which case he would stay on through Sunday, making sure he visited Spurgeon's Tabernacle to hear the famous preacher) he returned north.

Arrol was as ingenious as he was energetic. He had the ability to think and invent on the job. The riveting machines used on the Forth Bridge, for instance, were of his own design, developed specifically to overcome the unique problems of building this huge structure. Energetic, ingenious, pious and rising to fame and a knighthood from humble origins, William Arrol was, indeed, the very stuff of Victorian dreams.

He was born in Houston, Renfrewshire, the son of Thomas Arrol, a spinner. He went to work in a cotton mill at the age of nine and at 14 became a blacksmith's apprentice. In 1863, he joined Laidlaw and Son, builder and bridge manufacturers in Glasgow and, in 1868, launched his own business on his life savings of just £85. His hard work paid off. In 1872 he established the Dalmarnock Works and in 1875 undertook his first major construction, the building of the North British Railway bridge across the Clyde at Bothwell.

Arrol then went on to win the contract for Sir Thomas Bouch's abortive scheme for bridging the Forth and in 1882 became the successful tenderer for the greatest project of his career, Fowler and Baker's Forth Bridge. This undertaking – and the building of the new Tay Bridge – were described by the Prince of Wales when he opened the Forth Bridge in 1890 as 'monuments of his skill, resources and energy', and won him a knighthood that year. Meanwhile, in 1886, Arrol had begun work on London's Tower Bridge which was opened in 1894, once again by the Prince of Wales. Later contracts included the Nile Bridge at Cairo (1904-8) and the Wear Bridge at Sunderland (1805-9). In 1885 he bought the estate of Seafield, near Ayr, where he built a house, tended the gardens and collected pictures until his death in February 1913.

John Fowler 1817-1898
The Consulting Engineer

'The best of human race, who aye maintains a mild and kindly mood'

Born in Sheffield, John Fowler became one of the great consulting engineers of the age of Railwaymania, through which 'only men of iron constitution came unscathed'. He designed the Pimlico Bridge which carried the first railway across the Thames in 1860 and St Enoch's Station in Glasgow and was a pioneering engineer of the London Underground system.

'His history is the history of the most important period of the profession and dates back for more than half a century,' wrote Wilhelm Westhofen in 1890. On the death of Brunel, John Fowler became consulting engineer to the Great Western Railway. When he came to work on the Forth Bridge, Fowler had passed much of the responsibility for design to his younger colleague, Benjamin Baker. However, he remained the senior partner throughout the period of construction and was rewarded with a baronetcy in 1890.

Wilhelm Westhofen 1842-1925
The Bridge's Biographer

'It is no wonder that those engineers of the old school can turn from one subject to another with so much versatility when we consider what an education they had. Instead of having professors to fill them with ready digested knowledge like the young men of the present day, they were moved from one position of responsibility to another, and their intellects were hardened and invigorated by constant work. Every step they took was an experiment on a working scale, and every fact they learned was imprinted on their memories by the toil and trouble it had cost.'

In this extract from *The Forth Bridge*, Westhofen might well have been writing about himself. As the Bridge's 'biographer' as well as the supervising engineer of the Inchgarvie cantilever he demonstrated an outstanding ability to move 'from one position of responsibility to another'. Not yet 'of the old school' (he was only 48 years old when the Bridge was completed) he had considerable experience as assistant and draughtsman in Cologne and Mannheim before coming to

Wilhelm Westhofen

21

Taking soundings off Inchgarvie, 18 June 1884.
Evelyn Carey, the Forth Bridge's official photographer, was also an assistant engineer on the project. This arrestingly beautiful photograph from the series he took throughout the bridge's construction reveals his artistic eye as well as his professional understanding of the subject. The circular wooden raft was used to measure the contours of the rock surrounding the two south piers on Inchgarvie. It was strong enough to resist the action of the waves in an ordinary breeze and to bear the strain of being beached on shore in bad weather. The raft was a little under 70 ft. in diameter and carried four mooring blocks which were raised and lowered from four winches (centre). A sounding rod of fine steel wire with a 60 lb weight was wound round a drum slightly overhanging the raft (centre right). Two surveying instruments checked the position of the centre of the raft every few minutes and alterations in the tide level were observed and recorded on two tide gauges. Over a period, almost 3,000 soundings were taken in the hour before and after high water or low water. A diver swam round the raft, clearing away loose stones and picking off projecting sharp points.

London in 1867. There his work was varied and included investigating the manufacture of iron, steel and Portland cement, as well as experimental works for the proposed Channel Tunnel. In 1882 he was appointed Assistant Engineer for the Forth Bridge works, with responsibility for the foundations and building of the piers and the central, or Inchgarvie, cantilever. After the completion of the Bridge and the publication of his book, *The Forth Bridge*, Westhofen was appointed in 1891 to superintend the building of Gauritz River Bridge, the largest in the Cape Colony, to a design by Benjamin Baker. In August 1892 he became Head of the Engineering Branch of the Public Works Department of Cape Town.

Westhofen was elected Chairman of the South African Institution of Civil Engineers in 1906, but is equally well remembered in South Africa as a watercolourist of distinction. His work was frequently exhibited and he illustrated a book which became well known locally, Rene Juta's *The Cape Peninsula* (1910). Westhofen died in 1925.

The Forth Bridge first appeared on 28 February 1890 as a supplement to *Engineering* magazine which the highly respected James Dredge and William Henry Moore jointly edited. A repaginated version in book form was published in the same year and a centenary facsimile edition was published in 1989. An authoritative (and often entertaining) eye-witness account of the bridge's construction and an outstanding social history, *The Forth Bridge* provides a wonderfully complementary accompaniment to Evelyn Carey's photographic account.

Evelyn Carey 1858-1932
The Photographer

From the earliest days of photography, engineers recognised the potential of the new invention. In 1840 a Scotsman, Alexander Gordon, suggested to the Institution of Civil Engineers that photographs would enable 'views of buildings, works, or even of machinery when not in motion, to be taken with perfect accuracy in a very short space of time and with comparatively small expense'.

Evelyn Carey, the official photographer of the Forth Bridge was also an Assistant Engineer on the project. His work is outstanding in the history of industrial photography. Only rarely, even today, do photographers have the thorough and professional understanding of their subject and the aims of its creators that Evelyn Carey enjoyed. His remit was to record as accurately as possible the progress of construction as the structure rose over the Firth of Forth. This he carried out with technical accomplishment and precision, despite the great efforts which were required to handle his cumbersome equipment within the confines of space and light in the caissons and the steel mazes of the incomplete cantilevers.

Before Carey produced his glass plate negatives of the Forth Bridge under construction, the only comparable series to record the ambition and achievement of the Victorian engineers was Phillip Delamotte's work taken during the reassembling of the Crystal Palace at Sydenham, 1853-4. The haunting photographs of men working below sea-level in the caisson foundations which are reproduced in this book stand with the earliest attempts to capture hazardous working conditions and are contemporary with underground photographs of the mining industry.

THE FORTH BRID

ON SITE

As in a gigantic military manoeuvre, the bridge engineers brought men and materials to the site by specially erected landing stages. By the end of the first year, workmen's cottages on both sides of the Forth were filled to overflowing. Careful to recount the human as well as the technical side of the bridge building story, Westhofen noted the vast majority of the men were civilised and co-operative. A few were birds of passage who worked for a week or two and then passed on having contributed little to building progress but with full pay-packets in their pockets. 'But,' said Westhofen, 'black sheep are found everywhere . . . Taking them as a whole, it must be freely acknowledged that the workmen employed upon the bridge have not, to any material extent, added to the troubles and anxieties attendant upon such a work.'

The natural features offered by the chosen site were excellent: above both shores of the Firth, the land was sufficiently high to carry the railway lines onto the approach-viaducts of the Bridge while leaving adequate headroom for the largest naval and merchant ships; the water depth would allow the caissons, a vital element of the foundations, to be sunk; the rocky promontory of the Fife or north shore offered a bedrock for one of the cantilevers as well as anchorage for barges and launches connected with the construction work; the whinstone island of Inchgarvie overlaid with a bed of hard boulder clay (which Providence has so kindly placed in the middle of the Firth, said Westhofen) acted like a gigantic stepping stone on which to construct the central or Inchgarvie cantilever.

A more promising site was difficult to imagine and, as if to allay any lingering anxiety in a public over-conscious of the Tay Bridge disaster, Westhofen, who supervised the works at the Inchgarvie cantilever, wrote these reassuring words: 'There has been no single instance of the ground on which the foundations were placed being uncertain or in any way doubtful, and no anxiety need be felt in regard to this part of the Forth Bridge.'

On a visit to the site in 1883, Benjamin Baker remarked on the bustle and activity, both at the Queensferry Hawes Pier and on the north shore in contrast to the quietness, even dullness, which had once characterised these places. That year a cement store, general stores, cottages for the workmen, a large canteen and reading rooms were erected, and an observation building built at Port Edgar. The centre line of the bridge had been set out and the position of the masonry piers determined. Gas furnaces, engines, a large hydraulic accumulator, numerous cranes, drilling machines, a hydraulic crane and two large travelling cranes had been brought on site.

Opposite: North Queensferry, July 1883. The foundations for the north approach railway viaduct piers under construction as industry on a huge scale makes inroads through the peaceful sea-side hamlet. Inchgarvie island appears top left of Carey's photograph and new buildings associated with the bridge work stand on the shore opposite.

Below: The Forth Bridge site at South Queensferry, July 1883. The chasm from the south shore with the Hawes Inn and Hawes Pier in the foreground, Inchgarvie island with its ancient castle and new jetty for bridge works, centre, and the hamlet of North Queensferry on the opposite shore.

A wrought-iron landing-stage now stood on Inchgarvie, 'a narrow ledge of rock attacked by wind and waves from all sides', with the remains of a castle built around 1490 by James IV to protect shipping seeking refuge from pirates. Wind gauges had been erected in 1882 and, later, the keep of the castle, and outbuildings and battlements were roofed in for workshops, offices and stores to which a cottage, a kitchen and sleeping accommodation for 90 foreign workmen engaged in sinking the pneumatic caissons were added later. Air-compressing machinery, 576 holding-down bolts, 1,000 tons of steel plates and bars, 70,000 cubic feet of granite, were delivered to the sites where 38,000 cubic feet of concrete had been set and 5,000 yards laid. By the end of 1883 the four main Fife piers were in hand. The south-west Queensferry pier was well advanced and ready for the caisson.

The Scotsman reported that Fowler had built a comfortable dwelling house where, besides a handsomely furnished suite of apartments for his own occupation, there was accommodation for members of the resident staff. The engineers overlooked the bridge site and close to hand were an extensive model loft where patterns of actual working size could be handled, the immense workshop where the steel for the superstructure would be manipulated, and the yards and drill roads where the pieces of the world's longest bridge were assembled before being taken out to their final resting place.

That first year, machinery and plant of every description began to arrive on site and the pace was kept up at an astounding rate for the next five years at a cost not far short of £500,000. Much of it was specially designed or invented for the bridge works. Working at the frontiers of technology, William Arrol in particular demonstrated astonishing ingenuity in inventing on-the-spot items, including hydraulic spades for caisson work and a portable hydraulic riveter.

'Every step (the engineers) took was an experiment on a working scale,' said Westhofen, 'and every fact they learned was imprinted on their memories by the toil and trouble it had cost.'

A siding to bring in delivery wagons was laid from Queensferry Station, and another alongside the workshops. Cranes were set up in every convenient position so that sections of raw steel, plates, bars, angles and other hardware could be delivered quickly to the furnaces and drilling or planing machines in the Sheds.

The list of basic materials to build one of the wonders of the modern world read like this:

Steel – from the Welsh Landore Works, The Scottish Steel Company and Dalzell's Iron and Steel Works, Motherwell

Rivets – by the Clyde Rivet Company

Granite – from Aberdeen

Rubble – from Arbroath

Whinstone – quarried locally, or on site

Concrete – whinstone on Inchgarvie and at North Queensferry was exclusively used and broken down by stonebreakers and crushers

Sand – barges were sent 10 miles up the coast to Kinghorn and Pettycur; local proprietors objected to the removal of sand from the foreshore

Cement – exclusively Portland, manufactured on the Medway. An old hulk called the *Hougomont* brought it in and stored 10 or 12 tons at a time, moored off Queensferry. In 1886 the *Hougomont* was towed into Port Edgar after an outbreak of smallpox at Queensferry and converted into an isolation hospital

Timber – brought from Grangemouth in bulk as planks, battens and boards. Hardwoods including beech, oak and ash came from neighbouring sawmills

Creosote oil – for the Lucigen lamps and rivet and other furnaces

At the end of 1883, work was well under way to construct the north and south masonry piers, which would support the approach viaducts, and to build the cofferdams and caissons. The caissons for Inchgarvie and North Queensferry were already under construction on the beach at Queensferry. Arrangements were complete for the electric lighting of the whole works 'which not only [had] to be carried forward with the work, but also upward and in all directions . . . a task of considerable magnitude and of the greatest influence upon the rate of progress', as Wilhelm Westhofen, supervising the Inchgarvie cantilever, knew only too well. Gas rates at the time were 'ruinous' and electricity was used in the form of arc lamps for the shops and outdoor works and incandescent lamps for offices, stores and fitting shops. The arc lights were 1,500 to 2,000 candle-power but Westhofen complained of unsatisfactory changes in the colour and intensity of the light they gave and, when they failed, they put men in danger 'for while standing one moment in the dazzling glare of these lights they were sometimes suddenly called upon to use their eyes in absolute darkness or sit still'.

Lucigen lamps were used early on in the construction and 'for general work of erection they were the best for light, the simplest to keep in order and the easiest to attend to'. But leakages 'especially in high winds . . . covers girders and staging with a thick coat of slimy oil, making them slippery and unsafe'. At the height of construction the lighting arrangements required a separate department at the Forth Bridge Works where a large number of men attended to the lamps, the dynamo machines, the cables and other connections.

The lighting of the construction at various points and heights, which changed almost nightly in the later stages, caused confusion to shipping on the river. 'On such a night, with a slight mist on, the captain of a tug-boat, coming down river with a barque on tow, mistook the lights on the Fife erection for those of Inchgarvie, and steered his ship straight for the hamlet of North Queensferry which was hidden in the mist.' Eventually a lighthouse (which still stands today) was built at the north-west corner of Inchgarvie with a revolving light which was easily seen for twelve miles up and down river.

From the first year, nothing was left to chance. Men and materials were directed with all the vigour and technical prowess the engineers and supervisors could muster. With so many false starts well behind them and the promise of the completion of this colossal pioneering structure ahead – no structure of anywhere near comparable size had ever been contemplated before – the men who built the bridge moved from foundations to superstructure at an astonishing pace.

At the height of construction 4,600 men from Scotland, England, Ireland, Germany, France, Italy, Sweden and even Japan were employed on the bridge. Foreign workmen appeared in great numbers to lay the caisson foundations and, later, to construct the final stages which included laying the asphalt pavement for the permanent way, or railway line, across the bridge. 'Scots, English and Irish were about equally represented,' Westhofen noted, 'and though the latter furnished very few skilled hands, they were mostly very hard workers and very conscientious and reliable men.'

In *The Forth Bridge* Westhofen provides a lively account of social conditions during the seven years the bridge took to build. His attitude to the men is almost unfailingly caring, but his comments on strikers reveals him to be the boss he was:

Opposite: The Drill Roads, July 1885. The engineers' buildings overlooked the bridge site and included a huge model loft where patterns of bridge parts were handled to actual size, an immense workshop where steel for the super-structures was manipulated, and the yards and drill roads where the pieces of the world's longest bridge were assembled and disassembled before finally being taken out to be constructed as part of the bridge itself. John Fowler built a com-fortable house near the yards which The Scotsman *described as having 'a handsomely-furnished suite of apartments for his own occupation [and] accom-modation for members of the resident staff'.*

Below: Illustration of a drilling machine.

'Several strikes occurred during the building of the bridge, most of them brought about, not by the men themselves, but by organised committees in connection with various Trades Unions and their disputes with employers in other parts of Scotland. The causes were often trivial enough, such as the discharge from the works of some idle scamp with an inordinate allowance of the gift of the gab, and whose demand to be reinstated in his dignity at twenty-two shillings per week, caused an immense amount of useless suffering to scores of his fellow-workmen, and, more still to their families, and a proportionate increase in the takings of the neighbouring whisky shops.'

Within a year or two, accommodation for the briggers had reached bursting point and the North British Railway Company laid on special workers' trains from Edinburgh, Inverkeithing and Dunfermline. Later the Edinburgh line was extended to Leith where several hundred men chose to live in preference to the overcrowded conditions near the site even if that meant leaving Leith at 4 am and returning home at 7 pm.

On days when the weather was too bad for men to work on the bridge, the trains were turned back by telegraph, returning the men home rather than leaving them to hang about the Queensferry pubs. They spent a lot of time afloat too. At first a paddle-steamer was hired to take the briggers from shore to shore and to the foundation sites. Later it was replaced with a superior works vessel capable of carrying 450 men at a time, until gradually a fleet of steam barges, launches and rowing boats was built up. Two rowing boats were attached to each cantilever to act as rescue boats which, in the course of seven years' construction, saved eight lives as well as 8,000 workmen's caps and other articles of clothing blown or accidentally dropped off the bridge. The contractors supplied the men with boots and waterproofs during work on the foundations and added thick woollen jackets, overalls and waterproof shoes for superstructure work.

Large shelters and dining rooms heated by stoves were provided on the ground and on top of the central towers at viaduct level and at each end of the cantilevers as the superstructure developed. From the first summer the organisation of welfare for employees was impressive. A Sick and Accident Club was formed with compulsory membership at 8 pence a week for every person engaged by Tancred & Arrol which contributed 22 shillings a year.

'The wages paid to all classes of workmen were as a whole rather above the average,' asserts Westhofen, 'and as far the greatest amount of outside work was done by the piece, a skilful and steady workman was enabled to make double and treble his ordinary time wages if he applied his abilities and energies in the right direction.'

The Engineers' Drawing Loft, 4 August 1885.

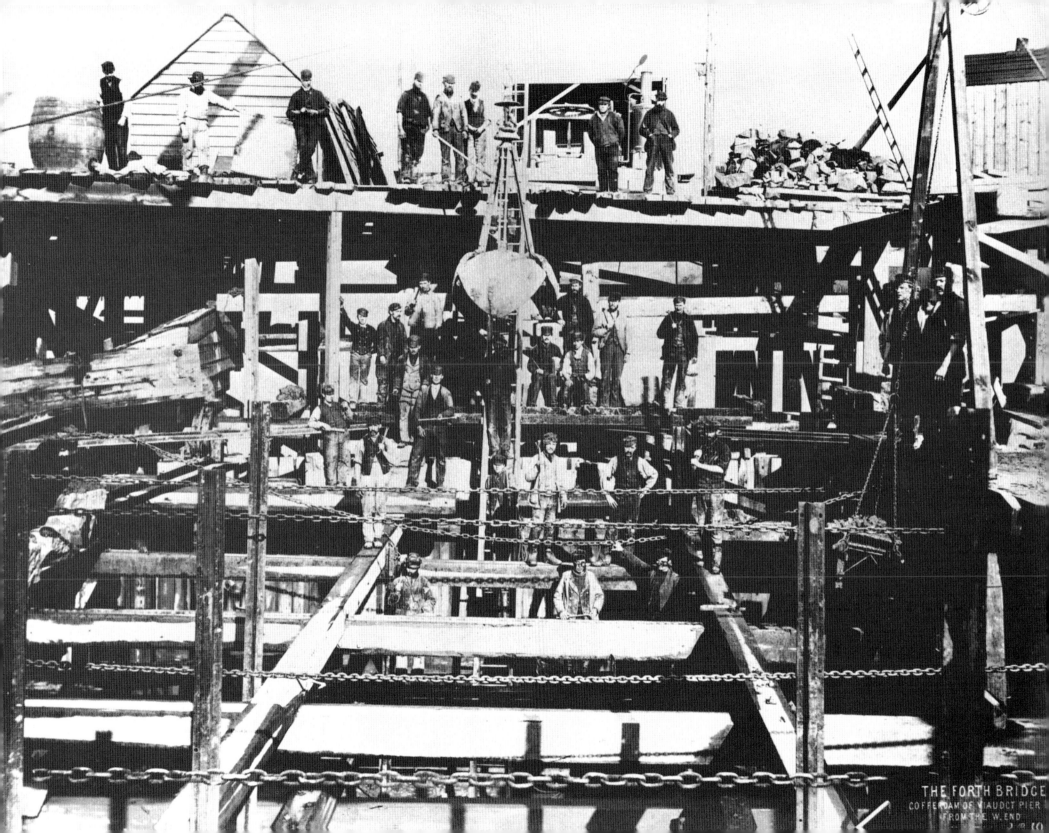

THE FORTH BRIDGE
COFFERDAM OF VIAUDCT PIER
FROM THE W. END

Page 31: Men at work on a viaduct pier, 4 April 1884.

Opposite: Queensferry cantilever foundation works, 4 April 1884.

Left: Workmen in one of the Inchgarvie foundations, 11 April 1884.
'No one need desire to have to do with a more civil or well-behaved lot of men, always ready to oblige,' wrote Westhofen. 'Scotch, English and Irish were equally represented as to numbers and though the latter furnished very few skilled hands, they were mostly very hard workers and very conscientious and reliable men.' Workmen also came from Germany, France, Italy, Belgium, Austria, Sweden and even Japan. Foreign workmen were employed in great numbers to lay the foundations and, later, to construct the railway line across the bridge. In Carey's photograph, water pours in to sink the caisson. The ladder on the left is the means of escape for the men.

Westhofen notes that caisson work required 'good health, freedom from pulmonary or gastric weakness, and abstemiousness, or, at any rate, moderation in taking strong spiritous liquors. Some of the most experienced hands of M. Coiseau [who advised on the caisson work] suffered when they had been making too free with the whisky overnight, and a good deal of the disorders that ensued were traceable to the same source; though, on the other hand, wet feet or incautious and sudden change from a heated atmosphere into a cold, biting east wind, insufficiency of clothing, and want of proper nourishment, had their influence in causing illness among the workers.'

Above: The Multiple Drilling Machine, 25
February 1887.
Opposite: The Mark Tapley bringing steel
into the works, December 1885.

THE FORTH BRIDGE.

BRINGING STEEL INTO THE WORKS.

DEC. 1885.

N 98

Opposite: Raising the north approach viaduct girders, 17 September 1885.
The human side of bridge building cheek-by-jowl with large-scale industry is touchingly captured by Carey's lens. A washing-day line of sheets, shirts and long-johns, hangs under the railway viaduct girders which were raised to 150 ft. above sea level at the same time as the masonry piers of the approach viaducts were built up underneath.

View from the south shore, 21 July, 1885. The approach viaducts have been raised on the south side masonry piers. The circular foundations for the Inchgarvie and Fife cantilevers are in place and Inchgarvie island appears to the right with the cliffs of North Queensferry beyond. A ramp appears in the immediate foreground with tracks for waggons which continue on to the jetty below, the means of transporting large quantities of material to the bridge site.

Riveters working on the drill roads,
12 March 1887

The Dart, *1885, was the engineers' and contractors' vessel.*

THE FORTH BRI
INCH GARVIE S.E. CA
May 15 1885 N.

FOUNDATIONS

It was impossible for the crowd which regularly gathered by the Hawes Pier to fathom how this bridge over the Forth would be constructed. There had never been a bridge like this before and it would be little short of a miracle if the bridge builders could pull it off. After the nightmare of the Tay Bridge disaster, would people risk travelling over it anyway?

In the meantime, though, there was a lot of noise and bustle, new sights to see and plenty to talk about. You could take a train out from Edinburgh and walk down to the beach at Queensferry where the enormous foundations were being constructed or up to the drill roads where all kinds of new-fangled equipment like electric lights were used. And you could touch the metal of the two temporary bridges linking the Edinburgh line with Queensferry which was salvaged from the fallen Tay Bridge and wonder at it all.

From the start there was the excitement of thousands of incomers arriving from all over Europe to live in North and South Queensferry, which within months had become hives of industry. There was a seemingly endless to-ing and fro-ing of trains and ships bringing in supplies. The first of many ceremonies took place to launch the first caisson, and there was an accident when one of the caissons tipped over and the water rushing out swept two briggers to their deaths.

The construction of the Forth Bridge fell, more or less, into two parts. The years 1882-1885 were devoted to sinking the caissons and building the piers which would support the superstructure. From 1886 to 1890 the superstructure itself was constructed, its progress visible for all to see from the shores of the Forth, as the Fife, Inchgarvie and Queensferry cantilevers rose higher and higher above the sea, finally to be joined by the massive girder arms.

Soon after contracts were signed, a start was made on temporary work at both North and South Queensferry.

Offices, stores, workshops and the No. 1 Shed, which had already been erected by William Arrol to back up work on Sir Thomas Bouch's ill-conceived suspension bridge, were retained and purchased by the new company. Preparatory work for the construction of the foundations included fixing sites for the masonry piers which would eventually support the great cantilever towers. The cantilever towers were designed to open out to a remarkable width of 120 feet at the masonry piers, so that the structure at railway-track level and above became light and compact in comparison.

Only 18 months after the contract had been signed, the south-west Queensferry caisson was launched on 26 May 1884 by the Countess of Aberdeen when a large crowd gathered to watch the first of several ceremonies which took place over the seven years' construction. Work was proceeding so quickly that Lord Aberdeen was back with his wife a year later to take part in the ceremony to launch the last caisson. After the contractor's representatives, Sir Thomas Tancred and Joseph Phillips, received the party, Lady Aberdeen was invited to turn the lever of the hydraulic launching machine which allowed the north-west caisson to glide down the slips into the water where two tugs towed it out to the site. A great cheer from the crowd accompanied this closing ceremony of the foundation works.

During 1885 two temporary bridges were built across the North British Railway line on the Queensferry side to connect with the Forth Bridge Works. The girders used in their construction fascinated the public. They had been salvaged from a portion of the ill-fated Tay Bridge. Offices were extended and drill roads added to. A new drawing loft was erected and the steel-work plant augmented. And as progress on the main masonry piers went forward at a pace, impromptu inventions were made to deal with the problems as

Inchgarvie south east caisson, 15 May 1885. Ceremonies were held to mark the launching of the caissons. The first caisson, the Queensferry south west, was launched in May 1884 by the Countess of Aberdeen. The last caisson, the south west Inchgarvie, was launched on 29 May 1885, again by Lady Aberdeen, who turned the lever of a hydraulic launching machine which sent the caisson gliding down the slips into the sea where two tugs waited to tow it to the site. 'A great cheer from the crowds accompanied this closing ceremony of the foundation works.'

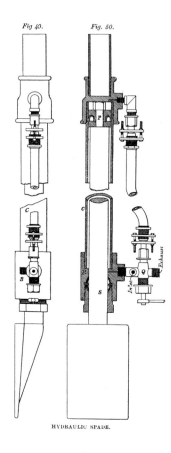

Fig 49.   Fig. 50.

HYDRAULIC SPADE.

Illustrations from Wilhelm Westhofen's
The Forth Bridge.
Above: William Arrol's hydraulic spade.
Right: sinking the Queensferry caisson.
Opposite: section showing permanent and
temporary caissons with airlocks and men
in the working chamber.

they arose. On one occasion, William Arrol visited an air chamber in a caisson to see what could be done to hurry up the work, and quickly invented a hydraulic spade.

On 21 December 1882, almost three years to the day after the Tay Bridge Disaster, the contract for the construction of the Forth Bridge was awarded to Tancred, Arrol & Co. by the Forth Bridge Company. Stringent safety margins incorporated into the design were tested on models by the engineers under different conditions, including the extreme of simulated hurricanes. Since a fatal failure of the Tay Bridge had been its inability to withstand excessive wind pressure, the tests on the Forth Bridge models were calculated to allow for maximum rigidity of the structure under the vertical stress of a moving train-load and under the horizontal stress of wind pressure. The Forth is particularly exposed to south-westerly gusts at the bridge site.

But Benjamin Baker, practical as ever, insisted on taking part in experiments not only on wind pressure, but also on

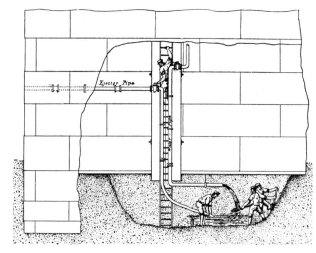

load pressure and tidal action. During the summer of 1882, wind pressure gauges were erected on top of the old castle on Inchgarvie and readings were taken from that time until the bridge's completion in 1890.

The designers put great effort into other important factors including ease and safety of erection, so that the unfinished structure was at all times during construction as invincible to the elements as the completed bridge would be. During erection, every stage of the work provided the secure base from which the next stage could proceed or a support from which temporary staging could be erected.

Before the contract was signed, the Board of Trade had insisted that the entire construction period be independently monitored to ensure the highest standards of workmanship and safety. Two engineers, Major Marindin and Major-General Hutchinson, were appointed to report quarterly from 1883 until February 1890. Together with the account written by Westhofen, they provide a wonderfully detailed report of month-by-month progress on the bridge.

The following summary of their report from 1883 to 1885 outlines the main work on the foundations:

1883: Offices and workshops at Queensferry are erected by June and equipped with machinery by August. Workshops and workmen's cottages also completed at North Queensferry. And at both places, as well as Inchgarvie, landing stages are built by August to enable men and materials to be brought in. Work begins on piers for both north- and south-approach viaducts, including building of cofferdams around the south-side pier-bases which would be under water at high tide.

1884: Queensferry Section: Caisson for south-west pier launched May 26. Both this and the south-east caisson in position by September, north-east caisson launched and north-west floated into position on December 4. Inchgarvie Section: North-east caisson in position by March and masonry work on this pier completed by year's end. North-west caisson lowered into position in September.

Fife (North Queensferry) Section: masonry work on north west and north-east piers complete by March and, on south-west pier, by December. Average number of men employed in building operation in previous quarter: 1,500 (June); 1,850 (December).

1885: Queensferry: North-west caisson tilts over and sinks on January 1. Recovery operations continue throughout much of year; not refloated until October 19. South-west pier complete by May, south-east by September and north-east shortly afterwards.

Inchgarvie: North-east and north-west piers complete and south-east caisson sunk by May. South-west caisson launched on May 29. Masonry work on south-east pier complete by December.

Fife: Timber staging between main piers to facilitate erection of horizontal tubes begun in May, and tubes in position by September.

Numbers employed: 2,800, including Italians engaged in caissons' compressed-air sinking arrangements (February); 2,200 (September); 2,000 (December).

THE CAISSONS

For all the ingenuity of its design, the magnificent super-structure of the Forth Bridge is only as stable as the groups of four platforms on which the cantilever towers stand; and where these platforms rest on underwater foundations, the key to success is in each case a 400-ton, 70-feet-diameter, wrought-iron cylinder known as a caisson.

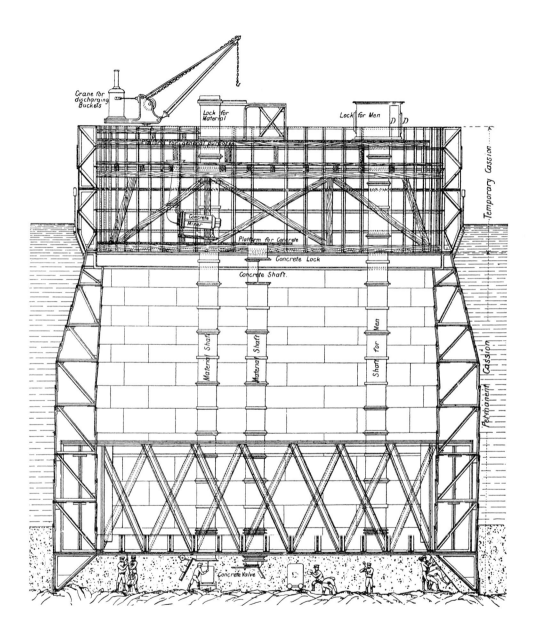

Wilhelm Westhofen described Arrol's hydraulic spade (page 42):
'Mr Arrol's ingenuity fortunately came to the rescue, and he devised a most simple, yet a most efficient, tool.' His hydraulic spade was capable of breaking up the hard boulder clay encountered on the sea bed which none of the existing tools could deal with. Evelyn Carey's eerie photographs show men using the hydraulic spade (pages 52 and 53). 'The spade was lifted by two men, a third attending to the fixing of the head and the turning of the cock admitting water pressure.' It must have been horrendously arduous work in the 7 ft. high chamber on the sea bed. 'The spade was set on the ground, the pressure turned on, and as soon as the head-piece had been firmly set against the ceiling the full pressure was given, and the spade forced down into the clay.'

Opposite: The Queensferry north east bedplate, 9 July 1886

The principle of the caisson is straightforward, but its operation in this case involved highly hazardous work on an epic scale. The six underwater caissons which were sunk pneumatically (two at Inchgarvie and four at Queensferry) were first constructed and put together by the firm of Arrol Brothers of Glasgow (there was no connection with William Arrol). They were taken apart again and sent to Queensferry to be re-assembled and riveted. The caissons for the Inchgarvie and Queensferry cantilevers were manufactured on the beach at Queensferry. The underwater caissons were floated out to the site of the future cantilever towers. Concrete was then poured into them until they sank to the estuary bed. A temporary dam was built around the top of each cylinder to facilitate work above the caisson while, on the sea bed, a steel cutting edge around the circumference extended some 7 feet below the concrete filled floor.

There was now only one way to secure the foundations of the caisson. The water in the chamber between the concrete and the river bed was pumped out and, in order to ensure that the pressure at that depth did not allow it to re-enter, it was replaced by compressed air. Shafts, each with an airlock to maintain the compressed atmosphere below, led to what now became a working chamber, 7 feet high. The airlocks were vital to the progress of work in the caissons and to the safety and comfort of the men.

Doors inside the airlocks led to the air shafts in the caissons. As soon as the air pressure on either side of a door was made equal (by the admission of compressed air) the air shaft opened to permit access down an iron ladder into the air chamber below. Workmen descended iron rungs leading into these shafts to secure a firm base for each caisson by literally digging away at the very foundations on which they stood. To return from the chamber to the outside, pressure within the airlocks was released into the atmosphere. The chamber was lit by electricity and the hydraulic spades invented by William Arrol were used to dig away earth and rock.

Once their efforts had sunk the caissons to the required depth, the working chambers were filled with concrete, masonry piers built above, the temporary dam around these piers removed, and the foundations made ready to receive the steel bedplates to which the superstructure would be secured.

'So perfectly was everything arranged in these caissons that even visitors were allowed to descend and inspect them,' observed *The Illustrated London News* in 1889. 'One day a number of salmon forced their way under the caisson and were, of course, secured. It was supposed that as these fish, when on their way from the sea to the river, always head against a current, they by chance had come upon the movement produced by the air escaping which is being constantly pumped into the caissons, had headed against it, and thus found their way as strange but not unwelcome visitors into one of the sights of the Forth Bridge.'

Some of the men working in the caissons suffered 'agonising pain in the joints, the elbows, shoulders and knee-caps, and other places [but obtained] instant relief on returning into the high pressure'. Westhofen does not dwell on the fact that working in the caissons must have been an appalling experience; he almost brushes aside the deaths of two men, 'both were already consumptive', who died during the caisson work, though not in the caissons themselves: 'the rigours of a Scotch winter had, at any rate, as much to do with their death as the air pressure. Another man became insane and had to be sent back to his own country.' It is not clear if the two men are the same as those who were drowned when the sea suddenly rushed into the north-west Queensferry caisson which ruptured as it was being pumped out.

THE FORTH BRIDGE
QNE BEDPLATE.

Carey's photographs document the construction of caissons on the beach at Queensferry.

This page: The foreshore with the lodge house leading to the Dalmeny Estate looks much the same today.

Opposite: The caissons of the Inchgarvie and North Queensferry caissons were constructed before being launched from Queensferry at high water during spring tides. 'As soon as afloat, a tug boat was attached and the floating monster at once towed to its final resting place, or else to the end of the jetty,' writes Westhofen. 'Should any tide not rise as much as was expected, it was the practice to hermetically seal the airshafts and other outlets from the working chamber and force air into the latter, in order to increase the buoyancy.' The first caisson was launched in May 1884 and the last in May 1885.

Page 48: North east caisson at Inchgarvie, 11 April 1884.
The man standing on the temporary staging, far right, may be Wilhelm Westhofen.

Page 49: Cofferdam of viaduct pier No. 7 from viaduct pier No.6, 4 April 1884.
Thirteen piers support the bridge's railway viaduct. The one shown here is on the north shore.

46

Above: North east caisson at Inchgarvie, 11 April 1884.

The group of visitors standing on the rubble masonry inside the caisson gives an idea of its scale.

The photograph below was taken on the same day as the one above, presumably shortly after the visitors had exited from the caisson up the ladder on the left. Soon after, the sluice gates were opened allowing the sea to pour in and sink the caisson.

Opposite: Queensferry north west caisson, 15 December, 1885.

Page 52: Queensferry north west caisson air chamber. Page 53: Inchgarvie south west caisson air chamber.

Evelyn Carey's haunting photographs of men working on the sea-bed foundations, taken in photography's infancy, are a remarkable achievement, though he himself was not entirely satisfied with the results. He explained the difficulties of taking photographs in artificially created conditions below water in an article in Industries, 20 April 1888. 'The views obtained though serving as records of the work carried out in the air chambers, have no pretentions to photographic excellence, definition and sharpness of outline being impossible to attain in the peculiar atmosphere at which the plates were exposed.'

Westhofen reports that 'photographs were taken in the [Inchgarvie] air chamber on several occasions, some of them requiring an exposure of fifteen to twenty minutes; but they were not very successful, owing to the changes in the atmosphere and the uncertain light of the arc lamps. Whenever the air pressure increased to a slight extent the atmosphere became quite clear and transparent; then the air would rush out at some point under the caisson edge with a

THE FORTH BRIDGE
QUEENSFERRY
N.W. CAISSON.
DEC. 15 1885.

noise like distant thunder and a great
wave of cold water came rushing back.
This caused a dense white fog to suddenly
rise in the air-chamber, which obscured
everything for a few moments and then
gradually disappeared again.'

The photograph on the right taken in the
Queensferry caisson, shows three men
working with William Arrol's ingenious
hydraulic spade.

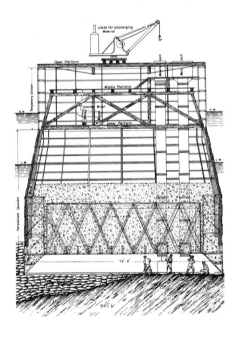

Opposite: Inchgarvie north east main pier, July 1884.
The granite pier foundation under construction above the sunk caission to support one of the skewbacks of the Inchgarvie cantilever. The array of 48 holding down bolts, in 4 rows of 12, each 25 ft. long and made of special steel, is well illustrated in this photograph.

Airlock for caissons, 6 May 1884 and illustration from The Forth Bridge.
The airlocks were vital to the progress of work in the caissons and to the safety and comfort of the men. Doors inside the airlocks led to the air shafts in the caissons. Air pressure on either side of a door was made equal (by the admission of compressed air) allowing the air shaft to open and permit access down an iron ladder into the air chamber of the caisson. To return from the air-chamber to the outside, pressure within the airlocks was released into the atmosphere.

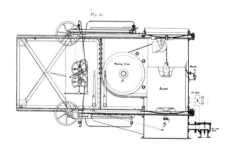

Viaduct girders at South Queensferry, 26 February 1885.

A small crowd watches progress on the construction of the girders for the railway viaduct which rests on the granite foundations of the masonry piers. There is a long way to go before it is lifted to a height of 150 ft. above sea level as the approach piers are built up underneath

Opposite: The Queensferry north-east cantilever bedplate, 9 July 1886.

The fixed lower bedplates consisted of four layers of steel, the first bolted into the granite foundation pier to a depth of 26 ft. and three layers above that. The fifth layer, a stiffening band round the edges of the bedplate, also retained the lubricant. Bedplates weighed between 33 and 44 tons each. The upper bedplates, similarly constructed, formed the base of the skewbacks, the lowest members of the cantilevers. 3,000 men were employed at this stage of construction.

THE FORTH BRIDGE.
Q.N.E. BEDPLATE.
Nº 126. JULY 9. 1886.

THE SACRIFICE

Fifty-seven lives were lost during the construction of the Forth Bridge. These deaths were included in the total figure of 518 men who were taken to the Edinburgh Hospital during the seven years of the bridge's construction. The fate of 461 injured men – one-tenth of the full workforce employed at the height of construction – is not recorded. Grim though this toll is, it came as a relief to the engineers and contractors who realised just how much worse it could have been on a structure which involved extensive work both below sea-level and at great heights. 'It is impossible to carry out a gigantic work without paying for it, not merely in money but in men's lives,' observed Benjamin Baker.

Wilhelm Westhofen noted that the use of wire ropes saved many lives. 'It may be considered a rash guess, though the writer at any rate has no hesitation in making it, that the list of accidents would have been doubled at least, and a great deal more time and money expended, but for the absolutely reliable character of [wire rope] . . . In the first instance there were nearly a score of cages or hoists for the raising of men and materials to the different levels at which work was carried out, and these were going continuously day and night for several years and there is not a single case on record of a rope having given way without having given ample warning . . . It is only necessary to call attention to the weight of these ropes in comparison with that of chain cables and hemp ropes of equal strength, to see at once the great advantages which the use of these ropes offered.'

The lack of accidents during the caisson operations was a remarkable achievement, given the danger of the work. The men in the caissons quite literally dug away at the foundations on which they stood. If the caisson sunk too quickly, they could all too easily be squashed between the sea bed below and the thousands of tons of concrete immediately above them.

Benjamin Baker had every reason to be apprehensive, for he was aware of the terrible fate of the workers in the caissons below the waters of the Neva at St Petersburg in 1876 when one caisson suddenly sank 18 inches. Of the 28 men in the chamber, nine remained imprisoned and only two were taken out alive. The rest were smothered in mud. He was also aware of the danger to men working in compressed air below sea-level from his knowledge of accidents in the previous decade, during the construction of New York's Brooklyn Bridge and the St Louis Bridge where 119 of the 600 employed in the caisson foundations work were affected by the 'bends'. Sixteen died and two were permanently crippled.

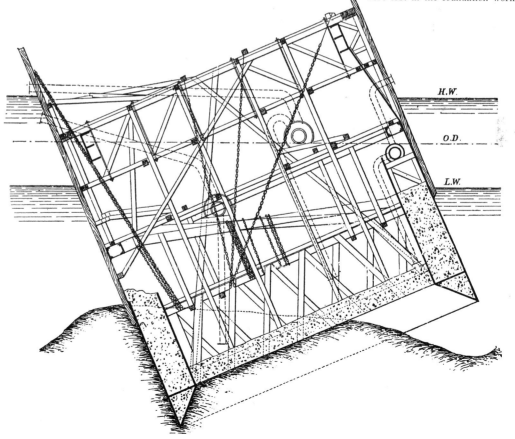

Tilted caisson at South Queensferry, January 1885.
'On New Year's Day, 1885, an exceptionally high tide occurred, followed by an equally exceptional low ebb, and the caisson sank deeply into the mud . . . Not being built high enough, the water soon flowed in, and filled it completely.' Reading between Westhofen's lines it is reasonable to surmise that the men were more intent on celebrating Hogmanay at the Hawes Inn than on bridge work. The caisson was not righted until October 1885. It was a remarkable achievement that no lives were lost in the foundation works.

H.W.

O.D.

L.W.

59

Inchgarvie, 6 April 1888.
The briggers spent a lot of time afloat. At first a paddle-steamer was hired to take them from shore to shore and to the bridge site. It was later replaced with a superior works vessel capable of ferrying 450 men at a time, and eventually a fleet of steam barges, launches and rowing boats was in operation. Two rowing boats were attached to each cantilever as rescue boats which, in the course of seven years' construction, saved eight lives as well as 8,000 workmen's caps and other articles of clothing, blown or accidently dropped off the bridge.

During the building of the Forth Bridge, not a single man died while working under the caissons on the sea bed – though there were some close shaves. On one occasion a few men were buried up to their chins in the mud and, on another, a caisson suddenly dropped 7 feet while men were working underneath.

The only deaths associated with the construction of the foundations occurred when the Queensferry north-west caisson ruptured while being pumped out. The sea rushed in and two men were killed. The story sometimes heard of another two men, trapped beyond rescue under a caisson, who were fed poisoned food to put them out of their misery is almost certainly apocryphal.

Working at a height was a different matter. It became impossible to avoid accidents. As the superstructure rose higher and higher, so did the number of deaths. The grim toll was coldly recorded in the Board of Trade Reports summarised below:

November 1887: 'Eight fatal accidents among the workmen during the past quarter, two having been killed during a gale of wind by a plank blown some distance from the North Queensferry pier. We were informed that none of the accidents was due to any failure of plant, such as has caused accidents in previous quarters.'

August 1888: 'Four fatal accidents during the quarter; one of these occurred to the engineer of one of the screw steamers, and is not attributable to the nature of the work; in another case, a rigger's help missed his footing and fell against a hand-rail which gave way. This was therefore a preventable accident as the hand-rail should have been more substantial.'

November 1888: 'Six deaths, but none appear to have been due to any want of care on the part of the contractors.'

August 1889: 'One fatal accident which we are informed was due to the workman's own want of caution.'

November 1889: 'We regret to report than even with [the] reduced number of workers, there have been four fatal accidents, one of which, on September 12th when a rivetter, John Aitken, aged 16, was killed, was due apparently to carelessness in the fixing of a stage upon which he was working. Two others, on the 18th and 24th September, are stated to have been due to the falling of pieces of wood from a height, which should not occur if proper care be taken. In the fourth case a rigger unfortunately lost his balance and fell into the water, receiving fatal injuries.'

The fearsome momentum gained by objects falling from heights on the bridge filled the engineers with trepidation. Baker reported having seen a hole straight through a 4-inch-thick timber which was made by a spanner dropped 300 feet. On another occasion, a spanner entered a brigger's waistcoat and exited through his trouser leg, ripping his clothes apart but leaving him uninjured. He also noted that a man who failed to close the rail on a hoist, stumbled and fell backwards a distance of 180 feet, 'carrying away a dozen rungs of a ladder with which he came in contact, as if they had been straws'.

An accident in the summer of 1887 brought about the only serious strike during the building of the bridge. When attempts were made to move a stage which had become jammed in a girder it collapsed, killing two men and a boy and injuring three others. The briggers downed tools for a week, demanding an extra penny an hour, which would have constituted a 15 to 20 per cent rise, as danger money.

'As might have been expected the principal spouters at the meetings held during the next following few days were men who worked in the yards away from all danger, or who did not work at all,' wrote Westhofen, 'and after holding out for a week most of the strikers were glad to be allowed to come back.' The supervisors had countered that most of the accidents which took place were the result of carelessness, the consequence, said Baker, of a familiarity with danger which bred a certain contempt.

Drink was another problem. 'The Hawes Inn flourishes too well for being in the middle of our works, its attractions prove irresistible for a large proportion of our 3,000 workmen. The accident ward adjoins the pretty garden with hawthorns, and many dead and injured men have been carried there, who would have escaped had it not been for the whisky of the Hawes Inn.'

Fife cantilever, south east skewback, 20 July 1887.
The skewbacks, a complicated assembly of five huge steel tubes attached to the foundations by bedplates, were of great importance as the element of construction which ultimately transferred the weight of the bridge's superstructure on to the granite foundations.

'From year to year the wonder grew, as the mighty piers slowly arose out of the sea and the ascending columns climbed ever higher and higher,' wrote Sir Robert Purvis in his *Memoir* of William Arrol. Crowds watched in wonder from the shore as the great cantilever towers and, later, the girder arms were constructed by men and machines suspended between sea and sky. 'More and more was the amazement as, week by week, the columns were perceived to be throwing out enormous, far-reaching growths on either side. Each of these was ever increasing in weight and altering in shape but ever in perfect balance.'

The first task ahead was to build the Queensferry, Inchgarvie and Fife cantilever columns. By the beginning of 1886 a large amount of material for the superstructure had already been prepared and every section was eventually built twice: 'The whole of the bridge is built in large sections in the works at Queensferry and the rivet fixtures tried so that no difficulty may be experienced in fitting up when the material is sent out to the piers,' reported *The Scotsman*. 'It thus comes about that the whole of the bridge will have been twice built ere it is finally erected.'

All the temporary work during the year consisted of making appliances to assist the erection of the cantilevers: hydraulic apparatus for riveting and lifting, specially designed machines for riveting the steel tubular members, Goliath cranes, hoists and steam-winders, compressors, accumulators, rams for high-pressure riveting, were all made ready. Three thousand men were employed now, and by the end of the year 33,000 tons of steel had been delivered the site.

The following summary from the reports to the Board of Trade for the years 1886 to 1889 outlines the rapidity with which the superstructure was constructed:

1886: Queensferry Cantilever: By August cranes, hoists, winches and rivetting machines delivered for erection of vertical columns to support cantilevers. Work on north-west caisson completed February 11. First skewback (south-east) in position by May; the next two (south-west and north-east) by August; and the final one (north-west) by November. Also by then, vertical columns at 116 feet. Inchgarvie Cantilever: Tubes between piers in position by March. Skewbacks completed by November.

Fife Cantilever: All four skewbacks in position by May and vertical columns at 171 feet by November. Numbers employed: 2,000 (March); 2,180 (May); 2,345 (August); 2,960 (November).

1887: The three vertical columns reach final height and building-out of cantilevers' members begins . . . after which, progress is measured in terms of distance reached by the 'branches' from their 'trunk'.

By November, the whole vertical superstructure above the piers is practically complete and lower members of cantilevers have grown thus – northwards and southwards respectively –

Illustration showing a skewback fixed to a granite foundation pier, a lifting platform which was raised up the structure by hydraulic rams and riveting cages where briggers worked. Each time a new section was riveted in place, the cage moved on. The Scotsman reported (March 1988) that 'every piece of work done becomes the basis of another advance, and the Forth Bridge men labour much in the same way as the Esquimaux who ascends the ice-cliff by cutting steps, one after another in its face'.

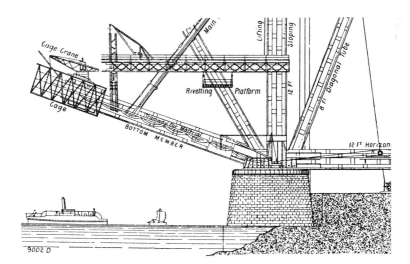

at south side: 73 and 60 feet (February); 130 and 122 feet (May); 152 and 126 feet (August); 160 and 152 feet (November). And at Fife side: 130 and 130 feet (May); 130 and 135 feet (August); 152 and 164 feet (November).

Numbers employed: 3,280 (February); 3,410 (May); 3,748 (August); 3,790 (November).

1888: At beginning of year, four double-derrick cranes are delivered which can be moved along the cantilevers' internal viaducts while helping with their construction. Girders there-fore begin to be connected in February on Queensferry, Fife and then Inchgarvie sections. As top members of each cantilever reach out and begin to converge with the lower members, they are connected by complex bracing system.

Numbers employed: 3,090 (February); 2,909 (May); 3,270 (August); 3,300 (November).

1889: Masonry work on approach viaducts almost finished by May: south end connected by temporary gangway to cantilever. All three cantilevers practically completed by August and by November central girders of two large spans are connected. All sleepers and rails in place on the south-approach viaduct and most of those on the north. Bridge can now be traversed from end to end.

The 12 solid foundation piers in three groups of four were finished off at the top with courses of granite masonry. The lowest part of each cantilever tower was constructed: steel plates held down by foundation bolts going down to a depth of 26 feet within the foundation piers; four layers of steel plates on each cantilever base, each weighing about 100 tons. The skewback was then fixed on each pier to hold together the 10 component members of the cantilever.

The skewbacks were the most complicated components of the entire bridge superstructure. As Mr Biggart, the engineer in charge of the drawing offices, shops and yards reported in a paper he read to the Institute of Engineers and Shipbuilders in Scotland a year before the bridge was finished: 'Each skewback is a starting point or a termination of five separate tubes and five distinct box girders . . . the difficulties [are] mostly on account of the various angles at which the different parts lie in relation to one another and the skewback.'

The engineers invented special appliances for riveting these skewbacks, which acted as the upper or sliding bed-plates, which bore some of the movement of the piers and cantilevers which was caused by expansion and contraction. The erection of the vertical cantilever columns was performed by means of a platform supported by the columns themselves and carried up with them, one on each side of the axis of the bridge.

In the workshops on each shore, drilling and fitting of materials for the internal viaduct, where the railway line, or permanent way, would be laid, was well in hand. By July, the Fife cantilever was well advanced and the north approach-viaduct piers had been carried up to within a few feet of their ultimate height. A start was made with building out the bottom tubular steel members of each of the cantilever arms at the same time as the cantilever towers were being constructed upwards. But the members of the top cantilever arms could not be started until the full height of each cantilever had been reached in 1887.

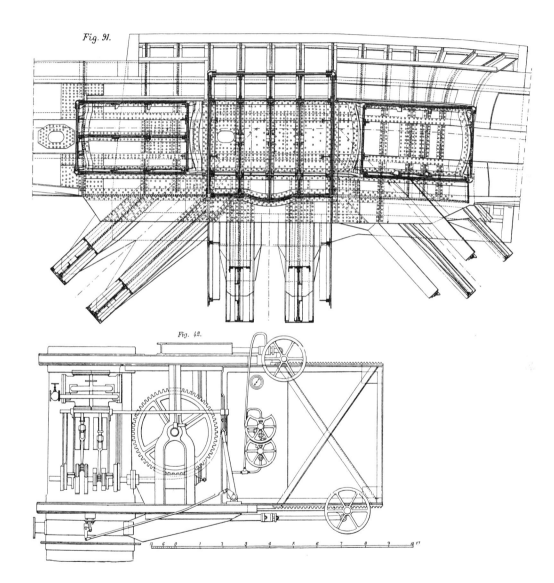

Fig. 91.

Fig. 42.

12 ft. tube, 5 August 1884.

Drilling a 12 ft. tube, 5 August 1884.

Everyone is fascinated by how the Forth Bridge is painted. When it was under construction, four firms were asked to tender as suppliers and samples of paint to the same specifications were tried out on the three cantilevers and the central girders. One of these firms, Craig & Rose of Leith, won the maintenance contract and has supplied its Forth Bridge Oxide of Iron Brushing Paint ever since.

As soon as they passed through the workshops or yards, every part of the superstructure – plates, bars, angles and tubes – received a good scraping down and a coat of boiled linseed oil. The insides of the cantilever tubes received one coat of red and two coats of white lead paint. Then, when erected as part of the bridge, they were treated with red lead paint. No less than 35,527 gallons of paint oils and 250 tons of paint were used during construction.

The early bridge painters, men fresh from the clipper ships, would shin up the exposed surfaces of the cantilevers, rope in one hand, paintbrush in the other. Today, the 16 bridge painters are hauled up and down on hydraulic lifts or cradles which meet Government safety requirements.

It takes three days and three nights to make a batch of Forth Bridge paint. But 100 gallons are always kept in store in case of demand, for the painters make the most of any sudden spells of dry and settled weather. Spirits, gel, synthetic resins, natural red oxide, barytes and talc are the magic ingredients which are blended in a huge ball mill (tile-lined and filled with pebbles and granite stones) kept specially for the purpose in Craig & Rose's Leith mills.

The painting area is estimated to be 145 acres, and 7,000 gallons would be needed to paint the bridge from end to end. Contrary to what practically everyone believes, it never is. To keep it up to scratch, areas most exposed to weathering are given priority, though it does get done over regularly in a rough cycle of four to six years.

Skewback under construction on the drill roads, 14 February 1885

67

THE FORTH BRIDGE.
THE FIFE CANTILEVER.
JUNE.7.1885. N.º 117.

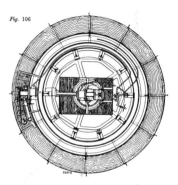

Fig. 106

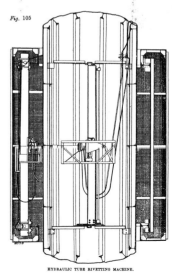

Fig. 105

HYDRAULIC TUBE RIVETING MACHINE.

Opposite: The Fife cantilever, 7 June 1886.

This page: Machine for riveting tubes, 17 June 1888.
A man at work in a circular wire cylinder placed round a riveting machine. 'Inside this cage the men worked with perfect safety as regards falling themselves or dropping things down on those working below,' notes Westhofen.

Diagram illustrating the erection of the cantilever towers with the lifting platform which hoisted steel from the vessels in the sea below, the caged riveting machines and the goliath cranes. The bridge acted as a scaffolding for its own construction since freestanding scaffolding could not be erected above water. 'As soon as a fresh round of steel plates is added to the tubes, or an additional girder section rivetted to the top arms, the platforms with their freight of men and cranes and other mechanical appliances are slid out correspondingly and a new piece of work is begun which again, when completed, will give the necessary standing support for a further extension.'

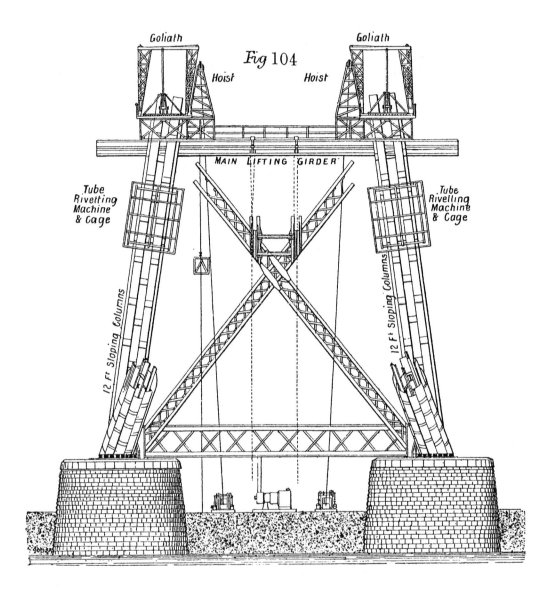

The Fife cantilever, 1 November 1888. The cage-like riveting machines moved up the tubes, keeping pace with construction. At least 6,500,000 rivets were used. Westhofen thought that figure was an underestimate: 'for on the Inchgarvie pier, where an exact record of rivetting was kept, the number closed there amounts to near upon 2,700,000 alone.' The lifting platform which received new material is carried aloft. Every part of the construction was tried out, drilled for riveting and prepared for painting on the drill roads before it was taken out to be erected as part of the bridge.

THE FORTH BRIDGE

THE FIFE CANTILEVER (LOOKING UP)

The three cantilever columns from the north side, 31 August 1887.
In Jubilee Year the cantilever columns, each with a section of the railway viaduct, reached their full height.

Opposite: The Queensferry cantilever, 12 March 1887.
It was possible for people to walk out to the Queensferry cantilever on the temporary jetty which ran parallel to the Hawes Pier (opposite). The hills of Fife beyond are covered in snow.

THE FORTH BRIDGE.
FROM N. SIDE AUG 5 1887.

Jubilee Year

The year Queen Victoria celebrated 50 years of golden rule, the three cantilever columns reached their final, majestic height. A reporter from the *Dundee Advertiser* was one of the visitors who climbed the cantilevers in Jubilee Year. There were two ways to go: you could reach the upper platforms by a staircase or by a cage rapidly drawn through the air by steam power. He chose the latter and reported: 'The cage is a small steel framework in which half-a-dozen persons may stand, but as it travels up through space, it resembles a canary cage. Once in it, visitors are as helpless as eggs in a basket and the sensation of being rapidly drawn through the air, up, up and still up, is peculiar. Ascending to the upper platforms, a magnificent view is obtained. A grand idea is also had of the stupendous nature of the Bridge works . . . From this upper platform one can gaze into the mouths of the great steel tubes. Their twelve feet of diameter look more, so massive and strong do they seem.'

Benjamin Baker thought the view from 'the summit' sublime. And Westhofen, standing on the top of the Inchgarvie cantilever which was his particular responsibility, wrote a poetic account of the vista:

'The broad river itself, with craft of all sorts and sizes, in steam or under sail, running before the wind, cutting across the current on the tack, or lazily drifting with the tide, is always a most impressive spectacle upon which one can gaze for hours with an admiring and untiring eye. And such it is, whether viewed in the glory of sunrise or sunset, in broad daylight with the cloud shadows flying over the surface, and a thousand ripples reflecting the sun's rays in every conceivable shade of colour, or in the soft haze of a moonlight night. The sunsets in summer are always magnificent, whether due to Krakatoan volcanic dust or to the vapours of the distant Atlantic, but there have also been many

HMS Devastation *sails past the Fife cantilever, March 1887.*

Opposite: 'Mist effect' recorded by Evelyn Carey's camera in 1887 from the south shore.

sunrises in early autumn when a hungry man could forget the hour of breakfast, and one could not find the heart to chide the worker who would lay down his tools to gaze into the bewildering masses of colour surrounding the rising light of day.

An unbounded view of more than 50 miles up and down river! Far away to the east the May Island, often so clearly defined, though 35 miles distant, that the sunlit cliffs are clearly visible, the Bass Rock and North Berwick Law, and the coast line of Haddingtonshire with the Lammermuir range fading into the sky, nearer Inchkeith with the white walls of the coastguard station and the lighthouse, Inchmickery and Cramond Island, the long jetties of Leith Harbour and the shorter of Granton and Newhaven, the full roads of shipping, the masses of houses in the marine suburbs of Edinburgh,

Arthur's Seat and Corstorphine Tower just peeping over Mons Hill and the woods of Dalmeny Park. To the south, the fertile districts of the Lothians gradually rising to the imposing range of the Pentland Hills, and to the south-west Dundas Hill and the Castle, Hopetoun House, and the old palace and church of Linlithgow, the harbours of Bo'ness and Grangemouth and the Campsie Hills closing in the upper Firth, still many miles wide with beautifully wooded shores and many towns and villages upon its banks.

Nearly 60 miles to the west, as the crow flies, stands the massive cone of Ben Lomond, and behind it a formidable array of hills and mountains, clothed in the summer-time in the tenderest shades of purple and blue, in the winter showing forth boldly in a coat of purest snow. In the north-west appears the Ochils, in the north and north-east the Fife

THE FORTH BRIDGE.

Lomonds and the beautiful coast of Fife running down into the horizon, where, glancing over the old priory on Inchcolm, the eye catches the May Island again.'

At the height of the construction in Victoria's Jubilee Year, more than 4,000 men were employed round the clock. During the night shift, the vast surrounding darkness was punctuated by the flashing lights of the island lighthouses, the harbour lights of Granton, Leith, Newhaven and Burntisland. Sometimes up to 200 vessels were anchored in the Firth, the long struggling lines of their masthead lights giving the appearance of a busy town having suddenly risen from the waters.

'On Jubilee night (21 June 1887), although the atmosphere was somewhat thick, 68 bonfires could be counted at one time on the surrounding hills and isolated points, while the great masses of the central towers of the bridge lighted up by hundreds of electric arc lights – Lucigen and other lamps – at various heights where work was carried on, formed with their long-drawn reflections in the waters of the Firth, three pillars of fire, and afforded a truly wonderful and unique spectacle.'

Crowds gathered on both shores of the Forth to celebrate. It was a turning point. With the construction of the three cantilever columns, the dream of the pioneering engineers was close to becoming reality.

THE FORTH BRIDGE.

FIFE MAIN PIER FROM COASTGUARD STATION.

Fife cantilever from the coastguard station, 20 December 1887.

A breathtaking view of the Fife cantilever under construction, showing a fabulous array of angles and patterns in cantilever arms, box and lattice girders and rectangular staging where the men stood to work on each piece of steel, step-by-step, up to the top member of the cantilever which is now in place. The bridge engineers rejected cast iron in favour of steel which was subjected to rigorous testing. (Cast iron and wrought iron was used in the ill-fated Tay Bridge and the Eiffel Tower was made of wrought iron). The Forth Bridge was the first steel bridge in the world. 42,000 tons of Scottish and Welsh steel were used in the construction of the cantilevers and girder arms: 'a more uniform, a more homogeneous, a more satisfactory material could not be wished for,' noted Westhofen.

Fife cantilever tower, 1887.
Industry's response to Nature, which had previously declared the chasm of the Forth unbridgeable, was on a massive scale, far beyond the dreams of most people watching in amazement as 'the mighty piers slowly arose out of the sea and the ascending columns climbed ever higher and higher,' and, in the words of Sir Robert Purvis, William Arrol's biographer: 'week by week, the columns were perceived to be throwing out enormous, far-reaching growths on either side. Each of these was ever increasing in weight and altering in shape but ever in perfect balance.'

Opposite: The Forth Bridge from the south shore, August 1887.
The three cantilever columns are complete and the railway viaduct dwarfs the buildings at the Hawes Pier. 'The Hawes Inn flourishes too well,' commented Benjamin Baker, 'for being in the middle of our works, its attractions prove irresistible for a large proportion of our 3,000 workmen. The accident ward adjoins the pretty garden with hawthorns, and many dead and injured men have been carried there.'

E FORTH BRIDGE.

Internal viaduct, 1888.
The cathedral-like proportions of the bridge above the railway viaduct are revealed in one of Carey's finest photographs. He makes the most of pattern and light in the acrobatic girders which seem almost delicate in contrast with the massive cantilever tubes soaring above. The mighty structure dwarfs the briggers and the group of boys on the left.

THE FORTH BRIDGE

The nation was avid for any news of the bridge's progress and the newspapers and periodicals of the day obliged readers with regular and extensive reports. Readers of *The Scotsman* learned about the process in which the bridge acted as a scaffolding for its own construction, since freestanding scaffolding could not be erected over the water for work on the cantilevers: 'The workers today are practically standing upon their labours of yesterday. As soon as a fresh round of steel plates is added to the tubes, or an additional girder section rivetted to the top arms, the platforms with their freight of men and cranes and other mechanical appliances are slid out correspondingly and a new piece of work is begun which again, when completed, will give the necessary standing support for a further extension . . . Every piece of work done becomes the basis of another advance, and the Forth Bridge men labour much in the same way as the Esquimaux who ascends the ice-cliff by cutting steps, one after another in its face'. (March 1888).

A reporter from *The Builder* described how the steel cantilever members had to be built out on either side of the cantilever towers at the same rate to hold the whole in equilibrium: 'The whole work, as we saw it from the engineers' steam launch, has a most remarkable appearance at present. The three piers (cantilever towers) suggesting the idea of three enormous ships with the tubes projecting like gigantic bowsprits over the water, and everything at present hanging out into the air unsupported.'

'And so night and day the work goes on, for when the sun goes down hundreds of electric lamps light up the great cantilevers,' reported the *Bradford Observer*, adding that 'in all but very stormy weather, the work may be said to be continuous, save on Sundays.'

Taking the story further, *The Scotsman* reported: 'At the ends of the cantilevers are large platforms above and below on which the men work . . . the upper platform carries two cranes, one for raising and lowering the pieces of steel into their places, and another that bears an immensely powerful hydraulic rivetting machine. This machine, which is not unlike a pair of shears, has been specially designed by Mr Arrol for the work. It exerts a pressure of forty tons and squeezes and welds the steel bolts as if they were made of clay.'

Throughout the construction of the cantilevers, the gigantic steel tubes which formed them were fitted together in Queensferry, taken out by works vessels and placed in situ by cranes, hoists and riveting machines. At this stage the men worked from cages which encircled and slid along the tubes to replace bolts with rivets. Finally, the gigantic projecting steel tubes were joined by extending a continuous girder between them.

Westhofen recounts: 'It was not long before the Jubilee cranes, which had originally started from the tops of the Inchgarvie tower, and had proceeded, the one towards the south and the other towards the north, met face to face the two others, which had come down from the tops of the Queensferry and Fife towers respectively, and with the putting in of the last lengths of booms, their functions practically came to an end.'

There was a gap of 350 feet between each cantilever after their arms, north and south, had been built to full length. The suspended girders which would now be attached were thus about 350 feet long and 40 feet high at the ends and 50 feet high in the centre. Exact measurements were made, taking into account anticipated expansion in the heat of the sun. Each girder was divided into eight slightly unequal lengths. Their erection on the superstructure, to close the gaps between the cantilevers, was a precise and highly organised operation. All the arms were finished by July 1889. The month before, the south arm of the Queensferry cantilever had been

Riveting the top member, 21 February 1889.

Two men and a boy on temporary wooden staging, suspended 360 ft. above the sea on a temporary platform, riveting the thickest pieces of metal on the bridge together, at the top of a cantilever. The Clyde Rivet company supplied about 4,200 tons of rivets for the superstructure. Each foot of the superstructure contains 100 rivets. Several types of drilling machines were in operation in the drill road sheds, all for special purposes. The multiple drilling machines could activate ten boring spindles at a time and move alongside the piece of metal being drilled. Work was carried on day and night with 'no deduction for meal hours during the night, the full twelve hours being paid for'.

so close to the approach-viaduct that planks could be laid across to allow the directors of the the Forth Bridge Railway Company to walk from the south shore to the northern arm of the Queensferry cantilever.

Efforts to complete the bridge reached fever pitch as the briggers on the Fife and Queensferry cantilevers vied with each other to become the first to cross to the Inchgarvie cantilever. Strong temporary platforms projecting some 25 feet beyond the end of the cantilevers were attached to the underside of the bottom members where the first half-bay of the suspended girder was erected. The girders were built out from the closing structure, half a bay at a time, each joint in the top and bottom boom being riveted before extra weight was added forward.

When the fifth bay had been added, the cantilevers were so narrow that the twin cranes which had done much of the lifting work could not be taken out further. The heavy platforms they carried were removed to make them lighter as they climbed up the curve of the girder arm towards the middle of the span. From here, the cranes lifted material from barges in mid-stream; and the heavy floor girders (4 tons each) were hoisted up by special tackle with wire ropes and steam-winches set up near the ends of the cantilever arms.

In September 1889 a wily brigger placed a ladder between the jibs of the cranes working on the cantilever arms and made a dangerous unofficial crossing 200 feet above water from the Queensferry cantilever to the Inchgarvie cantilever. By 10 October the gap had been structurally closed and on 15 October the workmen cheered Forth Bridge Company directors and their friends on the first authorised walk across to Fife. Only a 60 feet gap remained between the Inchgarvie and Fife cantilevers which had been temporarily bridged by a wooden structure for the occasion. Miss Constance Taylor, a niece of the Marquess of Tweeddale, Chairman of the North British Railway Company, was the first person across, escorted by the chairman of the Forth Bridge Railway Company. A few months later, her aunt, the Marchioness of Tweeddale, drove the first train across the bridge which transported some of the directors and various dignitaries.

The last booms were in place by the end of October and early in November the north-central girder was ready to be connected, a process which required a rise in temperature and consequent expansion of the girder to allow the key-plates to fall into place.

Westhofen was on hand to record the emotional moment of the bridge's completion: 'The temperature on that day did not rise sufficiently high to make the joint, but in the night a sudden rise took place, and by 7.30 in the morning the bottom booms were joined together for good. It now required a good fall of the temperature to get the top booms connected [because they were on a camber] . . . But the weather remained obstinate and it was not until the morning of November 14 that the key-plates could be driven in and the final connection made . . . And thus the Forth Bridge was completed – for the remaining work was simply to replace temporary connections by permanent ones, to rivet up those which were only bolted, and do the thousand and one things which always remain to be done after everything is said to be finished.'

'The thrilling portion of the story is done, and the novelist would wish to leave off with so dramatic an incident as that just told,' wrote Westhofen, revealing the creative side of his nature which was to emerge in the last years of his life, when he became a watercolour artist in South Africa. 'But there are yet some details which belong to the history of the bridge, and which could not very well be left unrecorded.'

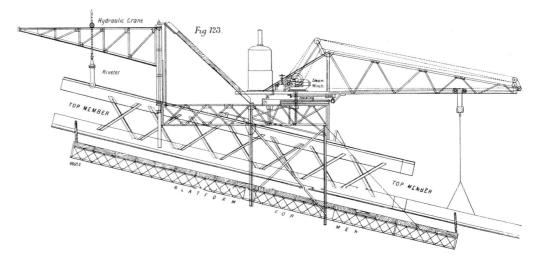

Junction at end of bottom member, Queensferry cantilever, February 1888. Skydiving box girder is pushed out by workmen from the temporary platform to link with the the huge skewback tube below. The fragility of the ladder is a reminder of the awesome danger the men faced working, often unprotected, at vast heights.

Fife cantilever, 24 May 1889.

Opposite: Queensferry north cantilever member, 13 March 1888.
Bird's eye view of Inchgarvie cantilever column and the island, right, from the summit of the Queensferry cantilever which was completed first. Benjamin Baker said the view from 'the summit' was 'sublime' and Westhofen described it in poetic terms: 'viewed in the glory of sunrise or sunset, in broad daylight with the cloud shadows flying over the surface, and a thousand ripples reflecting the sun's rays in every conceivable shade of colour, or in the soft haze of a moonlight night. The sunsets in summer are always magnificent, whether due to Krakatoan volcanic dust or to the vapours of the distant Atlantic. But there have also been many sunrises in early autumn when a hungry man could forget the hour of breakfast, and one could not find the heart to chide the worker who would lay down his tools to gaze into the bewildering masses of colour surrounding the rising light of day.'

Indeed, the completed Forth Bridge was subjected to stringent tests before it was finally declared open in March 1890. On 21 January two trains were taken on to the bridge. Their gross weight of 1,800 tons was made up of 100 coal wagons and six locomotives of 73 tons apiece. The trains were moved slowly abreast and halted three-quarters of the way through the central girder connecting the Queensferry and Inchgarvie cantilevers. Sir John Fowler and Benjamin Baker also took readings at other points along the bridge and were quite satisfied that the structure was adequately stiff.

This happy conclusion, combined with the bridge's easy withstanding of heavy gales before and after the loading tests, prompted the Board of Trade inspectors to report that 'this great undertaking, every part of which we have seen at different stages of its construction, is a wonderful example of thoroughly good workmanship with excellent materials, and both in its conception and execution is a credit to all who have been connected with it.'

THE FORTH BRIDGE.

THE FORTH BRIDGE
PLATFORM ON TOP OF QUEENSFERRY PIER

THE FORTH BRIDGE.

Opposite: Working platform on top of Queensferry cantilever, 13 March 1888. Many of the briggers were recruited from clipper ships; the group has a nautical air. On the 'deck' or top member of the cantilever is a bothy which contained a stove, and behind that, two hoists. In addition to steel for the permanent structure, hundreds of tons of weight were suspended from the cantilevers: 'cranes, temporary girders, winches, steam boilers, rivet furnaces, riveting machines, miles of steel wire ropes and of gangways, and acres of solid timber staging' is Westhofen's list. He added that 'a heavy shower of rain would in a few minutes put an extra weight of a hundred tons, and the storm would try its worst against these immense surfaces'.

Left: Fife north cantilever, 21 March 1888. Men work at the extremity of the structure. 'The workers today are practically standing upon their labours of yesterday,' reported The Scotsman in March 1888. The Fife cantilever stands on land where it was possible to build temporary wooden staging to provide a working platform.

Fife north cantilever from the coastguard station, 26 April 1888.
A sense of scale. Men work at every level. By 1888, efforts to complete the bridge reached fever pitch as the briggers on the Fife and Queensferry cantilevers vied with each other to become the first to cross to the Inchgarvie cantilever.

Fife north cantilever, 26 April, 1888. The mighty structure soars over streets and houses, the cantilever arms thrusting out to meet the railway viaduct supported by its masonry piers.

Illustration showing the erection of a cantilever, from The Forth Bridge.

Inchgarvie cantilever, 4 September 1888. The medieval castle and temporary buildings which provided canteens, offices and accommodation are dwarfed by the cantilever beyond. A bothy for the men stands on the railway viaduct. 'Large shelters and dining rooms heated by stoves were provided on the ground and on top of the [cantilevers], at viaduct level and at each end of the cantilevers as the superstructure developed.'

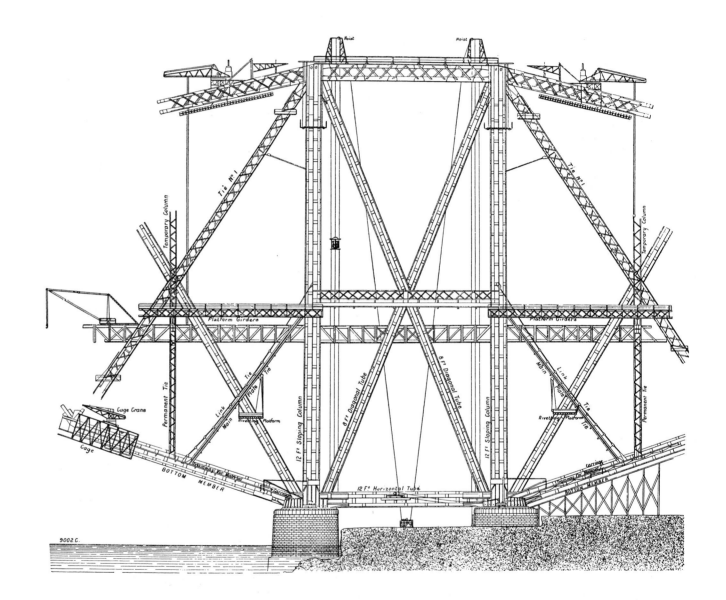

THE FORTH BRIDGE
CHGARVIE MAIN PIER . N. CANTILEVER.

THE FORTH BRIDGE
JUNCTION OF TIE AND STRUT

Opposite: Junction of the end strut, Queensferry cantilever, 16 February 1888. A brigger poses precariously at the extremity of the great bottom arm of the cantilever. On the right is a cage with a Jubilee crane which lifted steel for the next stage of the superstructure forward from a tramway behind it. William Arrol invented the crane which, like many artifacts of the era, was embellished with the name 'Jubilee' in 1887. A box girder tie, which runs all the way from the top of the cantilever tower (see illustration, page 90), is held by wire ropes to the junction it will soon be riveted to.

Left: 'It was not long before the Jubilee cranes, which had originally started from the tops of the Inchgarvie tower, and had proceeded, the one towards the south and the other towards the north, met face to face the two others, which had come down from the tops of the Queensferry and Fife [cantilever] towers respectively, and with the putting in of the last lengths of the booms, their functions practically came to an end.'

THE FORTH BRIDGE

Opposite: Riveting the top member of the Queensferry cantilever, June 1888. Westhofen recorded the process: 'The top member was rivetted in all parts by hydraulic machines wherever it was possible . . . two or three light timber stages followed at the back of the Jubilee crane, and here two or three rivet-heating furnaces were kept going to supply the various machines going below, the hot rivets being dropped down a long pipe, the end of which was stuck into a pail with ashes at the bottom.'

Left: the view under the Queensferry cantilever to the Inchgarvie cantilever, 15 April 1889, showing the gap to be closed by the linking girder. In September of that year a wily brigger placed a ladder between the jibs of the cranes on the cantilever arms and made a dangerous unofficial crossing 200 ft. above the sea from the Queensferry to the Inchgarvie cantilever.

Queensferry north west cantilever, June 1888, and the Forth Bridge near completion, 24 May 1889.
Riveters at work watched by workmates suspended on the end of a Jubilee crane. When the superstructure had reached this stage, light material, such as bars and angles were sent up the hoists while all the heavy booms and plates were placed on board one of the steam barges, and hoisted up by one of the twin cranes to the viaduct level to be taken by the Jubilee crane to the top. Later, when the top members had nearly approached the water, material for these was lifted by the Jubilee crane out of the barge and swung into place at once, saving time and labour. 'The erection of bays, 3, 4, 5, and 6 [sections of cantilever arms between tie junctions] was simply routine work . . . repetitions of the work gone through,' wrote Westhofen, 'and so much more easy for the reason that not only had the distances, both vertically and horizontally, between the members become so much less great, but also because the men had become so skillful and so accustomed to their tasks, that what appeared at one time to be insurmountable difficulties and hazardous undertakings, had now become mere child's play and were done in those exposed positions as easily as if the men were standing upon the floor of an ordinary workshop.'
Overpage: Illustration from The Forth Bridge showing the Queensferry and Inchgarvie cantilevers joined, and the gap between the Inchgarvie and Fife cantilevers still to be closed in 1889.

FORTH BRIDGE

THE FORTH BR
VIADUCT CIRDERS N

Approaching The Permanent Way

The first passenger train to use the approach roads and both railway lines on the bridge itself was driven by a woman, the Marchioness of Tweeddale, wife of the Chairman of the North British Railway Company. She epitomised the privileged Victorian woman who, vote or no vote, could achieve most things she set her mind on. To a reader in the late twentieth-century, saturated with media-hype, *The Illustrated London News* seems remarkably nonplussed in its report of the fact that she drove the train 'to and fro' on 24 January 1890 and that inside the train with her husband were the chairmen of the Great Northern and the Midland Railway Companies.

The internal viaduct which would carry the bridge railway lines (the permanent way), had been hoisted into the air with the mammoth bridge structure as it emerged, day by day, from the sea bed foundations, reaching its full height of 150 feet by 1888. The four rail troughs in which the double lines are laid are 18 inches deep and 16 inches wide and asphalted down to make a watertight base. Blocks of teak alternating with blocks of creosoted pine laid on the base were packed with a mixture of pitch, tar and black oil to set a hard foundation for the railway lines. Teakwood sleepers, held in position by wedge-shaped blocks of teak, were dressed to receive the rails. The total weight of rails, fish-plates, bolts, expansion joints for the double line of rails across the bridge was over 600 tons. Narrow footpaths on each side of the double line were made like asphalt pavements, with a slope to the outer side for water drainage.

Railway buffs, who are fortunate enough to obtain permission to walk over the bridge today, eagerly absorb information about the rails which are Forth Bridge Section, 56 feet long, weigh 125lbs per yard and are fastened to the timbers by coach screws. The main expansion joints on the rails are formed by long tongues cut at an angle of 1 in 63 and slide against a backing rail bent at a similar angle. A stepped plate arrangement always keeps the running rail to gauge.

The two direct railway connections to the bridge were constructed by the Forth Bridge Railway Company. The south approach railway extended from the south arches of the bridge to a junction with the station at Dalmeny; the north approach extended from the north arches to the station at Inverkeithing.

Subsidiary lines, on the south side to Corstorphine station and Winchburgh and on the north side to Burntisland, Townhill Junction, Kelty and Cowdenbeath, and from Kelty to Mawcarse through Glen Farg to a junction with the Bridge of Earn station, created a network giving access east, west and north.

Apart from the heavy cutting and tunnelling work, the greatest difficulty encountered in excavating the approach lines was a bog. Westhofen is uncharacteristically vague in his comment: 'Geologists will have it that this is the site of an extinct volcano, but it will probably be best to leave this question to be settled by geologists.'

The short south approach railway from Dalmeny was cut shale to an average depth of 20 feet. The north approach railway line from Inverkeithing was more complex with varying terrain over its two mile length. The work involved cutting through whinstone some 80 feet high, sinking shafts, creating covered ways and tunnels, creating embankments, viaducts and masonry piers. Pneumatic drills, dynamite and 'blasting powder' were the main tools used in excavating the tunnels and cuttings.

Viaduct girders on the north shore, 16 May 1885.
The viaduct, seen here near roof-top level with houses at North Queensferry, will eventually be raised to 150 ft. above sea level on the masonry piers. On the north side the approach roads extended to Inverkeithing.

Above: Briggers on a temporary platform within the viaduct girders, 17 September 1885.
Right: A cutting through whinstone rock on the north approach, February 1888. 'The Engineers with their gigantic works sweep everything before them in this Victorian era,' wrote Benjamin Baker of his colleagues who pitted themselves against every conceivable natural object to construct a bridge, a tunnel, a cutting or an embankment.

Opposite: Steam navvie with a digging bucket on the end of the jib forms a cutting for the railway and pulls a water bowser carrying coal to stoke the boiler, June 1888.

Opposite: North abutment arches, 23 June 1887.
Four masonry arches carried the railway viaduct beyond the masonry piers to the approach line. Here they are under construction on temporary timber centering.
Left: Men at work on a north approach embankment. Apart from heavy cutting and tunnelling work, the greatest difficulty encountered in excavating the north approach was a bog. 'Geologists will have it that this is the site of an extinct volcano,' noted Westhofen.

SITE OF THE CENTURY

The bridge site rapidly became a shrine to industry which any engineer worth his salt in Europe, Asia, Africa or America aspired to visit. Following the visit of the Prince and Princess of Wales in August 1884 scarcely a week passed without a visit from, as Westhofen put it, 'one-tenth of all people distinguished by rank or by scientific or social attainments'.

Visitors were actively encouraged on site in a way which is impossible to imagine in our ultra-safety-conscious society, and Benjamin Baker took great pleasure in the sport afforded by an ascent to the top of the towers he had designed. For him, it was all a great human experience as well as a technological triumph. In Jubilee Year he first demonstrated the cantilever principle using human models, which captured the imagination of people all over the world and which was endlessly reproduced in newspapers and periodicals of the day. He also travelled and lectured extensively, both at home and abroad, throughout the years of the bridge's construction. An imaginative and tireless educator, he relished capturing just the right analogy through which to communicate. In his 1887 lecture to the Edinburgh Literary Institute Baker suggested to his audience a way of grasping the enormous size of the Forth Bridge:

'To get an idea of the spans, let them stand on Waverley Bridge and look towards the Castle, and consider that the engineers had to span that distance with a complicated structure weighing 150,000 tons without the possibility of any intermediate pier or support; and let them consider also that their rail level would be as high above the sea as the Castle esplanade was above Princes Street, and that the steelwork of the bridge would soar 200 feet higher.'

For a London audience the same year, he translated the Edinburgh place-names to Green Park, Piccadilly and St. Paul's Cathedral.

The distinguished roll-call of international visitors included the Emperor of Brazil, the Kings of Saxony and Belgium, the King of Sweden, the Marquis Tseng of China and the Shah of Persia. *The Illustrated London News* reported that the Shah, His Majesty Nasr-ed-din, left the Highlands (where he visited the Balmoral estate and stayed as a guest at Invercauld House) on 22 July 1889 and travelled to Linlithgow to visit the Earl of Hopetoun. The following day he was shown the Forth Bridge and its workshops by Sir John Fowler and William Arrol after which he travelled by special train to Rothbury, Northumberland, to be entertained by Lord Armstrong, the munitions manufacturer, at Cragside.

Not content to view the wonder of the age from the shore, these 'hundreds of visitors, men of science of all nations, turbanned Indian princes and even venturesome young ladies', climbed all over the structure, and in the early years of construction, even into the caisson foundations.

'As in most other matters,' wrote Westhofen, 'ladies were to the fore, pluckily climbing into every nook and corner where anything interesting might be seen or learned, up the hoists and down the stairs and ladders, and frequently leaving the members of the so-called stronger sex far behind. It is, needless to say, that under such circumstances the duties of those called upon to guard the fair visitors were of the most agreeable.'

Opposite: Wilhelm Westhofen noted that the Forth Bridge site was visited by 'one-tenth of all people distinguished by rank or by scientific or social attainments'. The visit of the Shah of Persia in July 1889 was depicted in The Illustrated London News.

The Prince of Wales performed the ceremony of driving in the last rivet on 4 March 1890. His speech, which followed a sumptuous lunch, was recorded verbatim in The Illustrated London News the following week: 'It may perhaps interest you if I mention a few figures in connexion with the construction of the bridge. Its extreme length, including the approach viaduct, is 2,765 yards, one and one-fifth of a mile, and the actual length of the cantilever portion of the bridge is one mile and 20 yards. The weight of steel in it amounts to 51,000 tons and the extreme height of the steel structure above mean water level is over 370 ft., above the bottom of the deepest foundation 452 ft., while the rail level above high water is 156 ft. . . . about eight millions of rivets have been used in the bridge and 42 miles of bent plates used in the tubes, about the distance between Edinburgh and Glasgow. Two million pounds have been spent on the site in building the foundations and piers; in the erection of the superstructure; on labour in the preparation of steel, granite, masonry, timber and concrete; on tools, cranes, drills, and other machines required as plant; while about two and a-half millions has been the entire cost of the structure, of which £800,000 has been expended on plant and general charges. These figures will give you some idea of the magnitude of the work, and will assist you to realize the labour and anxiety which all those connected with it must have undergone. (Cheers). The works were commenced in April, 1883, and it is highly to the credit of every one engaged in the operation that a structure so stupendous and so exceptional in its character should have been completed within seven years. (Cheers).'

In a 'Forth Bridge Special' to celebrate the opening of the Forth Bridge, *Industries* magazine spoke for many Victorians when it expressed the hope that the achievement would help to bring about a dramatic shrinking of the earth's surface and bring its inhabitants yet closer together. The Forth Bridge was a great Victorian dream-come-true, a triumph which chimed with 'Land of Hope and Glory' and Empire-building. An illustration of the train called 'Progress' which appeared on souvenir programmes of the opening ceremony on the 4 March 1890 displays a similar aspiration on its hoarding which declares a route around-the-world, 'Aberdeen to New York via Tay Bridge, Forth Bridge, Channel Tunnel and Alaska' and is captioned 'Of dazzling great adventures this the foremost'.

The Illustrated London News reported that The Prince of Wales (later Edward VII) fulfilled his task 'in a strong breeze of wind, on that lofty open structure, necessarily in a hasty manner, but with vigorous goodwill and cheerfulness, in the presence of an illustrious assembly, comprising his son Prince George, who had come with him from London to Edinburgh the day before; His Royal Highness the Duke of Edinburgh, who had travelled from Russia on purpose; the Duke of Fife; the Earl of Rosebery, who was the host of their Royal Highnesses at Dalmeny; the Duke of Buccleuch, the Marquess of Tweeddale, the engineers and contractors of the Forth Bridge, some directors of the four railway companies interested in this undertaking; the Earl of Elgin, the Earl of Wemyss, Lord Colville of Culross, Lord Balfour of Burleigh, Lord Polwarth.' There were people from all walks of British life and many representatives of foreign railway companies and engineering works, including M. Eiffel, M. Picquard and other prominent scientists of the day.

The train went slowly over the bridge to North Queensferry allowing the party to inspect the huge cathedral-like structure and to appreciate the breathtaking view of the Firth of Forth and its shores. They embarked at the North Queensferry pier on board the steam-launch *Dolphin* and sailed under one span of the bridge, round the isle of Inchgarvie and to the south side, returning to the north pier. The two Princes and their party then re-entered the train, which moved back over the bridge. In the middle of the north connecting girder the train stopped to allow the Prince of Wales to perform the ceremony of driving the last rivet.

A hydraulic riveter was swung from one of the booms, the gilded rivet placed in the bolt-hole and the silver key to screw it in place handed to his Royal Highness by Lord Tweeddale. The Prince of Wales, with Mr Arrol's assistance, finished the work in a few seconds amid cheers. The rivet is on the outside of the railing and holds three metal plates together. Around its gilded top there is an inscription stating that it is the 'last rivet driven in by his Royal Highness the Prince of Wales, 4th March 1890.'

The train stopped a second time at another platform on the south cantilever pier where several women stood. The wind was blowing so violently that his Royal Highness had difficulty in retaining a steady foothold; it was impossible to make a speech. He simply said: 'Ladies and Gentlemen, I now declare the Forth Bridge open.'

The engineers' enormous model loft at Queensferry was transformed into a banqueting hall for the splendid lunch and honours which followed. Strips of white calico edged with blue and red concealed the wooden roof and a magnificent canopy of crimson and gold velvet was decorated with the Royal Coat of Arms and the motto of the Prince of Wales on a gold scroll.

Shields bearing the arms of British towns and the arms of the railway companies also embellished the building which for the previous seven years had been solely devoted to utilitarian purposes. 'All Reap at Last the Actions they have Sown'

Above: The Engineers' Model Loft was transformed for the opening ceremony banquet.

Opposite: The back cover of the lavishly designed menu, toasts and honours list which was presented to guests at the banquet.

declared the souvenir menu and on this great occasion the engineers and contractors could truly enjoy the fruits of their labour and accept honours bestowed.

William Thompson, chairman of the Forth Bridge and Midland Railway Company, and John Fowler were made baronets and Benjamin Baker and William Arrol were knighted in the first of many accolades which were to come their way at home and abroad in the years that followed.

People were reluctant to travel over the bridge at first and the habit soon arose of throwing coins from the carriages into the sea to ward off any lingering spectres of the Tay Bridge disaster and bring luck and a safe crossing to those who did.

The Forth Bridge was the ~~~ ttraction for tourists
dinburgh Exhibition
adows and featured
, a music hall and a

s. The historian and
is loathing of it and
t specimen of all
rwise: 'The Eiffel
roportioned and of
ge is a work of
s very ugly, but it
construction of a

lack of artistic
yal Academician
ler: 'One feature
ornament. Any
vould have been
e is a style unto

Fancy might try, in vain, to paint upon the page of thought,

Thy image; for thy mighty form with loftiest grace is fraught,

Each charm, which from our ken was hid in unseen worlds concealed,

Here, gathered in thy perfect self, is to our eyes revealed.~

Souvenir programme cover. The hoarding of a train called 'Progress' announced a route around-the-world: 'Aberdeen to New York via Tay Bridge, Forth Bridge, Channel Tunnel and Alaska'.

itself: the simple directness of purpose with which it does its work is splendid and invests your vast monument with a kind of beauty of its own, differing though it certainly does from all the beautiful things I have ever seen.'

In our own time Sir Kenneth Clark firmly set the Forth Bridge in the tradition of the great achievements of mankind. A sweeping view from the top of a cantilever was chosen to illustrate the cover of his book *Civilization*. Throughout the twentieth century the image of the bridge has been used countless times to advertise everything from shortbread tins to ladies' stockings, longlife batteries and micro-electronics – the symbol par excellence of strength, reliability and permanence.

The wild Forth estuary had been harnessed by the engineers of the Railway Age but within a few decades a new motoring public was dreaming another dream of not having to wait in frequent bad weather for the ferry boats to take them from shore to shore or to drive the long way round via Kincardine. People wanted a road bridge which would whisk them swiftly over the Forth as if on a magic carpet. With the opening of the Forth Road Bridge in 1964 the Forth's second bridge of dreams came true.

There are many other road bridges like it all over the world. But 'The Bridge' is unique. Every year, thousands of people come to gaze at Scotland's superbridge. And for Scots themselves, the Forth Bridge is like a childhood friend, part of life's reassuring landscape, a treasure to be cherished in its second century.